U. Winkler   W. Rüger   W. Wackernagel

Bacterial, Phage and Molecular Genetics

An Experimental Course

Springer-Verlag
Berlin Heidelberg New York 1976

Professor Dr. Ulrich Winkler
Doz. Dr. Wolfgang Rüger
Priv.-Doz. Dr. Wilfried Wackernagel

Ruhr-Universität Bochum
Abt. für Biologie, Lehrstuhl für Biologie der Mikroorganismen
4630 Bochum-Querenburg, FRG

Translated by G. Schulte-Hiltrop and W. Rüger

The first edition of this book was published in German also by Springer-Verlag under the title "Bakterien-, Phagen- und Molekulargenetik", XI, 285 p., 1972.

ISBN 3-540-07602-6 Springer-Verlag Berlin Heidelberg New York
ISBN 0-387-07602-6 Springer-Verlag New York Heidelberg Berlin

Library of Congress Cataloging in Publication Data. Winkler, Ulrich, 1929– . Bacterial, phage, and molecular genetics. Translation of Bakterien-, Phagen- und Molekulargenetik. Includes bibliographies and index.
1. Bacterial genetics-Experiments. 2. Viral genetics-Experiments. 3. Molecular genetics-Experiments. I. Rüger, Wolfgang, 1936– joint author. II. Wackernagel, Wilfried, 1941– joint author. III. Title. QH434.W5613. 1976. 576'.1. 76-8394.
Offsetprinting and bookbinding: Julius Beltz, Hemsbach/Bergstr.

Preface

During the mid-forties bacteria and phages were discovered to be suitable objects for the study of genetics. Genetic phenomena such as mutation and recombination, which had already been known in eukaryotes for
a long time, were now shown to exist in bacteria and
phages as well. New phenomena as lysogeny and transduction were discovered, which gained great importance
beyond the field of microbial genetics.

Bacteria and phages are of small size, multiply rapid-
ly, and have chemically defined growth requirements.
Many selective procedures can be applied to screen for
rarely occurring mutants or recombinants. Therefore,
they offered ideal conditions to investigate genetic
processes and to interpret them in molecular terms.
Many new methods were developed (e.g. CsCl density
gradient centrifugation) and old techniques were improved and modified for new purposes (e.g. chemical
mutagenesis). Hypotheses, such as the semiconservative
replication of DNA, mutation by transition and trans-
version and operon regulation, have had an extraordi-
narily stimulating effect on the research in general
genetics. Thus, in the past two decades, from the genetics of microbes (including fungi) the field of molecular genetics developed. Many text books compete in
presenting the latest knowledge on this subject. But
to date there are only a few laboratory manuals which
introduce the student of biology to the manifold experimental techniques of microbial and molecular genetics. This laboratory manual is an attempt to redress
the balance.

The experiments were selected for this book such that
each part of microbial and molecular genetics is about
equally represented. Care was taken to ensure that most
of the experiments could be performed with standard
laboratory equipment. Among the experimental objects
used here, are *Escherichia coli* and some of its phages
and also *Serratia marcescens* and phage κ. At present,
certain interesting experiments (extracellular UV
mutagenesis) can be carried out more conveniently with
Serratia phages than with *E. coli* phages; in other
experiments, such as phage crosses and complementation,
it is less important which particular phage is used.
Obviously a study of phage crosses could be extended
to circular linkage maps, different types of heterozygotes and the use of deletion mutants for mapping,
in which case phage T4 would be more suitable.

The experiments described were tested in several laboratory courses lasting from 3 to 6 weeks. Each time approximately 20 students of biology and biochemistry participated. The students were in about their third year of study, and up to this time had had little practice in experimentation. In order to facilitate experimentation by the beginner and to help him to find his way through the multitude of test tubes on his bench, we have taken care to give detailed descriptions of the experimental procedures. We have also included extensive data sheets for the execution and evaluation of the experiments. On the one hand, these data sheets are to serve as example for the recording of experimental results; and on the other hand, they are to aid the instructor to criticize the work of 20 or more students in as short a time as possible. The students can compare their own records and results with the data obtained in our laboratory while working out the experiments for this course (Section IV). Our results of the experiments should not be considered as absolute. Changes in laboratory conditions (chemicals, equipment) naturally influence the results to a certain degree.

At the beginning of each experiment, we have briefly summarized the theory on which it is based. These introductions can not substitute for a text book; but, with the literature quoted at the end of each experiment, they should help the student to brush up on his theoretical knowledge and, thus, to gain insight into the experiment.

We have written this book to stimulate the student's interest in genetics by giving him detailed instructions to "rediscover" experimentally genetic phenomena which he already knows. Besides this we wanted to make the experiments quantitatively evaluable and to demonstrate a series of methods of experimental and theoretical-statistical nature.

Requests for the necessary bacteria and phage strains (Section III/B) used are welcome. Specimens will be sent at a minimal charge.

Acknowledgements. At this point we would like to thank all technicians of the institute who have helped us in working out the experiments. We are indepted to Mrs. Sigrid Mickley who has typewritten the English manuscript and to Mrs. Rosi Winkler who prepared the graphs. We are also grateful to Professor R.W. Kaplan for his valuable advice on the statistics section. We wish to express our thanks to all our colleagues who placed strains of bacteria and phages at our disposal.

Bochum, January 1976   The Authors

Contents

VIII

I. General

A. Abbreviations and Expressions Frequently Used

1. Nucleic Acids and Bases

A	adenine	DNA	deoxyribonucleic acid
T	thymine	RNA	ribonucleic acid
G	guanine	mRNA	messenger RNA
C	cytosine	tRNA	transfer RNA
U	uracil	rRNA	ribosomal RNA

2. Units

length $\quad$ $1 \text{ m} = 10^3 \text{ mm} = 10^6 \text{ μm} = 10^9 \text{ nm} = 10^{10} \text{ Å}$

weight $\quad$ $1 \text{ g} = 10^3 \text{ mg} = 10^6 \text{ μg}$

volume $\quad$ $1 \text{ l} = 10^3 \text{ ml} = 10^6 \text{ μl}$

radioactivity $1\text{Ci} = 10^3\text{mCi} = 10^6\text{μCi}$ (Ci = Curie)

$$1 \text{ Ci: } 3.7 \times 10^{10} \text{ radioactive decays per sec or}$$
$$2.2 \times 10^{12} \text{ radioactive decays per min}$$

3. Other Abbreviations and Symbols

Abbreviations for the characterization of the genotype of bacteria and phages are listed in Section IIIb. For the abbreviations of nutrient media and solutions see Section IIIa.

cpm	counts per min
dpm	decays per min
ε	molar extinction coefficient
EDTA	ethyldiaminetetraacetic acid
η	refractive index
x g	gravitational force
k, -k	growth constant or inactivation constant
log cells	bacteria in the logarithmic (exponential) growth phase
M	molar
m	multiplicity of infection = phages/bacterium
μ	mutation rate, induced
N	total number of cells, viable cells or plaque formers

O.D.	optical density, synonymous with extinction
Φ	quantum yield
P-buffer	phosphate buffer
ρ	density (g/ml)
S	Svedberg unit
σ	effective cross section
SSC	standard saline citrate
stat cells	bacteria in the stationary phase
t	time
T_m	melting point
rpm	revolutions per min
UV	ultraviolet light

We have tried to be consistent with the lettering most frequent-ly used in other textbooks. Thus, it could not always be avoided that one symbol was used for two different meanings. Deviations from the definitions listed above are given in the text of the respective experiment.

4. Explanation for Some Expressions Used in the Text

- Denaturation of DNA. The conformational change of native DNA by physical or chemical treatment, i.e. the melting of double stranded DNA (helix-coil transition).

- Fractioning a gradient. The emptying of a centrifuge tube, fraction by fraction, at the end of a CsCl or sucrose gradient centrifugation. Usually the bottom of the tube is pierced with a hypodermic needle and a constant number of drops per fraction is collected.

- Soft-agar layer plates. Agar plates, layered with 3 ml of soft agar, which contains 0.1 ml of a bacterial suspension and, when needed, the addition of 0.1 ml of a phage suspension. Often used for the determination of viable counts or plaque titers.

- Indicator bacteria. When determining phage titers by the plaque method, indicator bacteria form the bacterial "lawn" on which the plaques are formed.

- Culture, logarithmic. A culture of growing bacteria, such as one obtains by a 2-4 hrs incubation of a 1:50 dilution of a stationary culture. Log cells are freshly grown prior to the beginning of the experiment and then kept in an ice bath until they are used.

- Culture, stationary. A bacterial culture which has been grown over night (approx. 15 hrs) and whose cells are in the sta-tionary phase.

- Culture, rolling. A bacterial culture which is aerated by rolling in a slanting position during incubation. This procedure prevents the formation of foam!

- Lysate. A suspension of phages which have been released by lysis of infected bacteria.

- Marker, genetic. A mutation, by which a gene is "marked" for genetic experiments, e.g., for crosses. Point mutations must make the mutants phenotypically distinguishable from the wild type, if they are to serve as genetic markers.

- Marker, selective. A mutation, which under suitable conditions, provides the mutated organism with a selective survival or growth advantage.

- Multiplicity of infection. The ratio of infecting phages to bacteria in the infection mixture.

- Scintillation counting. A method which allows a quantitative measurement of the decay of radioactive substances by means of the transfer of the energy of radioactive decay to scintillators (organic ring compounds). The scintillations emitted by the scintillators upon excitation by radioactive decay are then registered electronically.

- Streaked plates. Agar plates upon whose surface 0.1 or 0.2 ml of a bacterial suspension has been evenly spread with a sterile glass rod. The method is used to determine the number of colony forming bacteria (viable counts).

- Centrifugation, low speed. Centrifugal forces up to 5,000 times the gravitational force are employed. Practicable at room temperature.

- Centrifugation, high speed. Centrifugal forces up to 50,000 times gravitational force are employed. The rotor must be cooled to compensate for heating due to friction.

- Centrifugation, ultra. Centrifugal forces higher than 50,000 times gravitational force are employed. The rotor must spin under vacuum and must be cooled in order to avoid aerodynamic buoyancy and frictional heating.

B. Basic Equipment for the Experiments

The experiments were planned for groups of two students each.
The following items of laboratory equipment are used in all
the experiments. Hence they are not listed again in the "Material"
section of each experiment.

1. Material for each Group

- 1 plastic bucket for used pipettes
- 1 plastic tub for used test tubes
- 1 stop watch for laboratory use
- 1 hand tally for counting colonies or plaques
- 1 bunsen burner with tubing and lighter
- 1 pipette aid, e.g., propipette
- 2 pairs of goggles
- 1 wooden block used as a stand for 2 inoculation loops
- 1 pair of pointed forceps
- 1 large-holed test tube rack, empty
- 1 can each with sterile 10, 1 and 0.1 ml pipettes
- 1 styrofoam container for ice
- 1 beaker (250 ml) to hold two glass rods, suitable for streaking
- 1 glass rod with fine polished ends, 4-5 mm diameter
- 1 crayon
- 1 pocket lens, approx. 6x magnification
- 1 screw-cap bottle with approx. 200 ml sterile phosphate buffer
- 1 wash-bottle with distilled water

2. Material for two Groups

- 1 counter for the enumeration of colonies and plaque plates
- 1 large-holed test tube rack, 2/3 filled with large sterile test tubes, and to 1/3 filled with small sterile test tubes (a total of 60-90)
- 1 box of soft absorbent paper for the cleaning of cuvettes, etc., e.g., Kleenex

1 dissecting microscope (magnification 5 to 10x)

1 water bath at 47°C with small-holed test tube rack (for soft agar layer tubes)

1 water bath at 30° or 37°C

1 spectrophotometer (e.g. the Bausch and Lomb (SPECTRONIC 20) with cuvette holder and 2 round cuvettes (diameter 1 cm)

1 microscope with phase contrast optics (magnification 10 × 40)

1 Petroff-Hauser counting chamber

C. Calculation of Titers and Some Statistical Methods

All studies in quantitative biology are made with the goal of obtaining results which are fairly representative. However, in most experiments one studies only a small sample. In order to make this procedure valid, samples should be taken randomly.

1. The concentration (titer) of colony-forming cells in a given bacterial suspension can be obtained by taking a random sample from the suspension, diluting it appropriately and spreading 0.1 ml of an appropriate dilution on nutrient agar. After over-night incubation the titer can be calculated from the colony-count by dividing the number of colonies found by the factor of dilution. Details of the calculation may be taken from the example given below.

Dilution	Sample	Colonies/0.1 ml	Statistical weight (arbitrary)
1×10^{-4}	1	560	1
	2	625	1
3×10^{-5}	1	138	0.3
	2	171	0.3
1×10^{-5}	1	45	0.1
		1,539	2.7

Titer: $1,539/2.7 \times (1 \times 10^{-4}) = 5.70 \times 10^6/0.1$ ml
$\underline{ 5.70 \times 10^7/ml}$

Bacterial titers calculated in this way become increasingly reliable by enlarging the size of the sample. The size of the sample is taken into account by the square root of the total number of colonies considered $(\pm \sqrt{N})$. The ratio $100/\sqrt{N}$ (%) = v expresses the error as percentage of the titer calculated. For more information see e.g. CAVALLI-SFORZA, 1969, p. 49 ff.

$\sqrt{1,539}/2.7 \times (1 \times 10^{-4}) = 39/2.7 \times 10^{-4} = 0.15 \times 10^7/ml$
$\underline{\text{Titer} = 5.70 \pm 0.15 \times 10^7/ml}$

$v = 100/\sqrt{1,539} = \underline{2.6\%} = 0.15/5.70$

The same procedure is also applicable to the determination
- of phage titers based on plaque counts on indicator plates.
- of cell titers based on microscopic cell counts.

2. Significance tests. If two or more random samples are taken from a larger group, the composition of the group from which each sample was taken can be predicted by the use of statistical methods. In order to determine whether two such groups are composed identically (e.g. whether two bacterial cultures contain an identical number of a certain mutant) the samples can be compared by a "significance test". By a similar statistical procedure one can decide whether the experimental results match the results predicted by a hypothesis.

a) Using the t- or STUDENT-test, one can calculate whether the difference between two measurements $(x_1; x_2)$ is significant or not. The t-test can only be applied if the values to be compared represent random samples from two normal distributions of nearly the same variance. Instead of absolute values one may also use the arithmetic means $(\overline{x}_1; \overline{x}_2)$ from each of a series of continuously or discontinuously varying values. (After CAVALLI-SFORZA, 1969.)

$$t = \frac{\overline{x}_1 - \overline{x}_2}{s_D} \times \sqrt{\frac{z_1 \times z_2}{z_1 + z_2}} \qquad (1)$$

$\overline{x}$ = the arithmetic mean from several measurements (absolute frequencies), e.g. average number of colonies per plate

z = number of samples taken, e.g. number of plates

$z - 1$ = number of degrees of freedom

s_D = standard deviation of the difference $(\overline{x}_1 - \overline{x}_2)$

$$s_D = \sqrt{\frac{\Sigma (x_{i1} - \overline{x}_1)^2 + \Sigma (x_{i2} - \overline{x}_2)^2}{z_1 + z_2 - 2}} \qquad (2)$$

For t-values calculated according to Eq. (1), the corresponding P-value (level of significance) can then be taken from Figs. 2 and 3 (PÄTAU, 1943). For details see (2d).

b) The Chi-test can be used to investigate whether the difference between two empirically found percentage frequencies is significant or not. This test is only for discontinuous variables. Chi is equal to the difference between two single frequencies $(p_1; p_2)$, divided by the square root of the sum of the squares of the standard deviations $(s_1^2; s_2^2)$:

$$Chi = \frac{p_1 - p_2}{\sqrt{s_1^2 + s_2^2}} \qquad (3)$$

$p = M/N \pm s$

M = sum of elements of interest in a random sample, e.g., recombinants, mutants, etc.

N = sum of all elements in the same sample, e.g., mutants and non mutants.

s = standard deviation = $\sqrt{p (1 - p)/N}$ $\qquad (4)$

Standard deviations can be read directly from Fig. 1 (KOLLER, 1940). The procedure is as follows:

-   If a point on the p-scale is joined by a straight line to a point on the N-scale, the middle scale will be intersected at the corresponding value s or 3s, resp. If this intersecting line rises less than the next neighboring "licence line", when observed from the N-scale, then the sample size N is too small for the frequency p. For example, when p = 3.0% then N must be ⩾500.

-   If p is larger than 50% then the reciprocal value must be inserted.

For all Chi-values the corresponding P-values (level of significance) can be taken from Figs. 4 and 5 (PÄTAU, 1942).

c) With the Chi^2-test one can investigate whether absolute or relative frequencies of several events found in random samples (observed frequencies) deviate significantly from "theoretical" frequencies which are expected on the basis of a given hypothesis. The experiments No. 12 and 13 of this book include Chi^2-tests. P-values can be taken from Figs. 4, 5 and 6.

d) The level of significance, P, is a measure of the significance of the difference between an observation and a hypothesis. Consequently, only small P-values (⩽0.01) are really valuable, i.e. those indicating that the observed data do not agree with a given hypothesis. If P-values, however, are greater than 0.05 the hypothesis under consideration is neither excluded nor proven. The meaning of P can be best illustrated by an example: Let us say a certain experiment was performed once and a P-value of 0.003 was found. This means that the chance of again finding the same or a greater deviation between observation and hypothesis is 0.003 : 1 or 1 : 333.

Generally the following limits of significance are accepted:

P ⩽ 0.01	the difference is significant
P = 0.01 to 0.05	the difference is probable
P > 0.05	the difference is uncertain.

Literature

CAMPBELL, R.C.: Statistics for Biologists, 385 p. Cambridge: University Press 1974.

CAVALLI-SFORZA, L.: Biometrie. Grundzüge biologisch-medizinischer Statistik, 211 p. Stuttgart: Fischer 1969.

DOERFFEL, K.: Beurteilung von Analysenverfahren und -ergebnissen, 98 p. Berlin-Heidelberg-New York: Springer 1965.

KOLLER, S.: Allgemeine statistische Methoden in speziellem Blick auf die menschliche Erblehre. In: Handbuch der Erbbiologie des Menschen (Hrsg. G. JUST). Berlin: Springer 1940.

KOLLER, S.: Neue graphische Tafeln zur Beurteilung statistischer Zahlen, 167 p. Darmstadt: Steinkopff 1969.

KLECZKOWSKI, A.: Experimental Design and Statistical Methods of Assay, p. 616-730. In: Methods of Virology, Vol. 4. London: Academic Press 1968.

MATHER, K.: Statistical Analysis in Biology, 267 p. London: Chapman and Hall 1973.

PÄTAU, K.: Eine neue Chi^2-Tafel. Z. f. Induktive Abstammungslehre 80, 558-564 (1942).

PÄTAU, K.: Zur statistischen Beurteilung von Messungsreihen. Biol. Zentralblatt 63, 152-168 (1943).

SPIEGEL, M.R.: Theory and Problems of Statistics. Schaum's Outline Series, 359 p. New York: McGraw-Hill 1972.

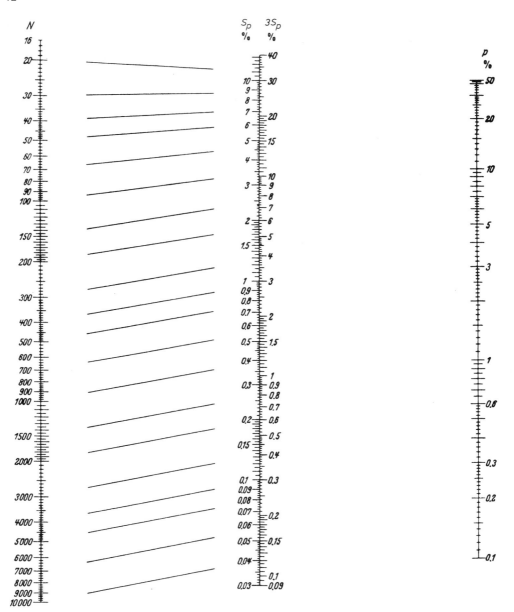

Fig. 1. Nomogram for the standard deviation s (in %) of frequency p (in %) for N observations. From KOLLER, 1940, Illus. 20

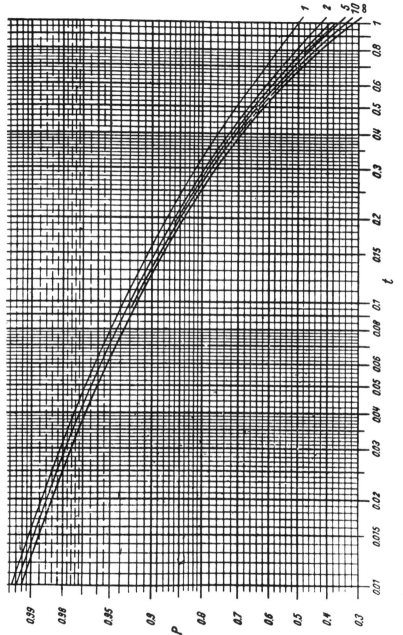

Fig. 2. t-Table for the determination of the level of significance P for the range
t = 0.01 to t = 1. From PÄTAU, 1943, Illus. 1

14

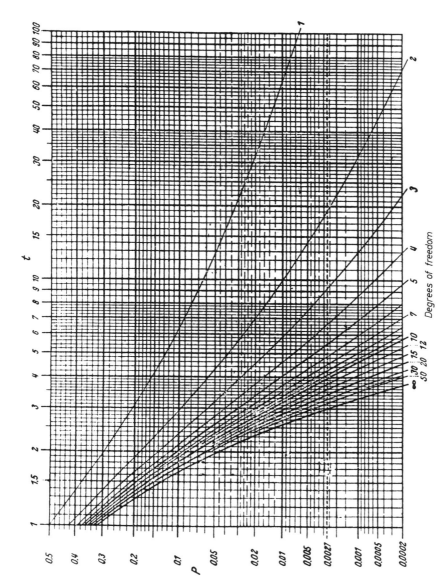

Fig. 3. t-Table for the determination of the level of significance P for the range t = 1 to t = 100. From PÄTAU, 1943, Illus. 2

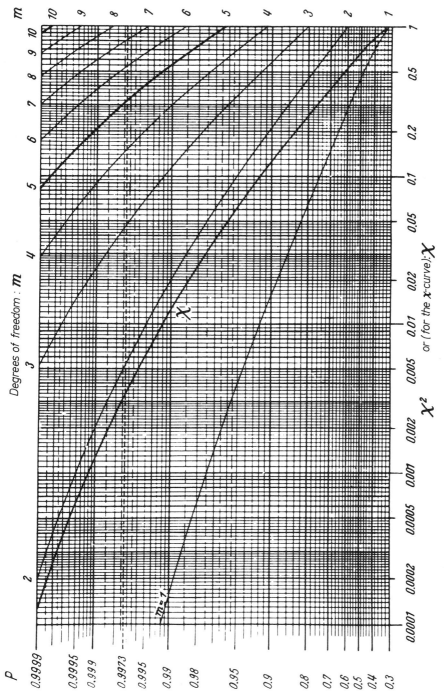

Fig. 4. Chi²-Table for the determination of the level of significance P for the range Chi² = 0.0001 to 1. From PÄTAU, 1942, Table I. The degrees of freedom (m) correspond to z−1 in the text (p. 9)

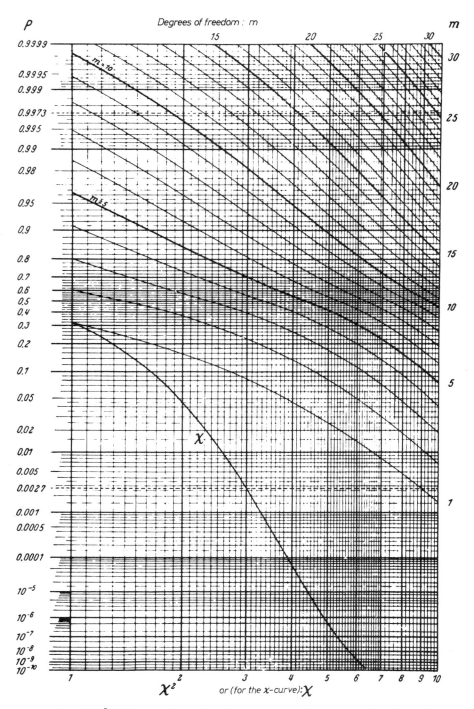

Fig. 5. Chi^2-Table for the determination of the level of significance P for the range Chi^2 = 1 to 10. From PÄTAU, 1942, Table
II. The degrees of freedom (m) correspond to z-1 in the text
(p. 9)

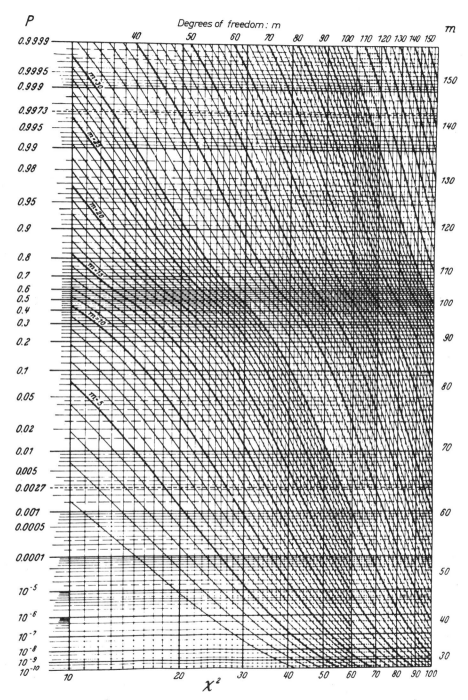

Fig. 6. Chi²-Table for the determination of the level of significance P for the range Chi² = 10 to 100. From PÄTAU, 1942, Table
III. The degrees of freedom (m) correspond to z-1 in the text
(p. 9)

D. Technical Literature

Textbooks

BRESCH, C., HAUSMANN, R.: Klassische und molekulare Genetik, 415 p. Berlin-Heidelberg-New York: Springer 1972.

CLOWES, R.C., HAYES, W. (eds.): Experiments in Microbial Genetics, 244 p. Oxford: Blackwell Scient. Publ. 1968.

HAYES, W.: The Genetics of Bacteria and their Viruses, 925 p. Oxford: Blackwell Scient. Publ. 1968.

KING, R.C.: Handbook of Genetics, Vol. 1. Bacteria, Bacteriophages and Fungi, 676 p. New York: Plenum 1974.

MILLER, J.H.: Experiments in Molecular Genetics, 466 p. Cold Spring Harbor: Cold Spring Harbor Laboratory 1972.

STENT, G.S.: Molecular Genetics, 650 p. San Francisco: Freeman Co. 1971.

WATSON, J.D.: Molecular Biology of the Gene, 662 p. New York: W. Benjamin, Inc. 1970.

Collection of Selected Articles

Many of these articles are suitable for short seminars which can provide practice in summarizing original publications.

ADELBERG, E.A. (ed.): Papers on Bacterial Genetics, 450 p. Boston: Little Brown 1966.

HAYNES, R.H., HANAWALT, P.C. (eds.): The Molecular Basis of Life (Readings from Scientific American), 368 p. San Francisco: Freeman Co. 1968.

SRB, A.M., OWEN, R.D., EDGAR, R.S. (eds.): Facets of Genetics (Readings from Scientific American), 354 p. San Francisco: Freeman Co. 1969.

TAYLOR, J.H. (ed.): Selected Papers on Molecular Genetics, 649 p. New York: Academic Press 1965.

ZUBAY, G.L., MARMUR, J. (eds.): Papers in Biochemical Genetics, Second Edition, 622 p. New York: Holt, Rinehart and Winston, Inc. 1973.

ZUBAY, G.L. (ed.): Papers in Biochemical Genetics, 554 p. New York: Holt, Rinehart and Winston, Inc. 1968.

Reviews and Symposia

Most of the following titles appear annually. The articles report progress in the field of microbial and molecular genetics and therefore fairly good knowledge is expected.

Advances in Genetics. London: Academic Press.

Annual Reviews of Genetics. Palo Alto, Calif.

Annual Reviews of Biochemistry. Palo Alto, Calif.

Annual Reviews of Microbiology. Palo Alto, Calif.

Cold Spring Harbor Symposia on Quantitative Biology. Cold Spring Harbor, N.Y.

Fortschritte der Botanik (Progress in Botany). Berlin-Heidelberg-New York: Springer.

Progress in Nucleic Acid Research and Molecular Biology. London: Academic Press.

Textbooks of Biochemistry

DAWSON, R.M., ELLIOTT, D.C., JONES, K.M. (eds.): Data for Biochemical Research, 654 p. (a handbook). Oxford: University Press 1969.

LEHNINGER, A.L.: Biochemistry, 1104 p. New York: Worth Publ. Inc. 1975.

MAHLER, H.R., CORDES, E.H.: Biological Chemistry, 1009 p. London: Harper & Row Publ. 1971.

II. Experiments and Problems

A. Phage Growth and Ultracentrifugation

For many experiments in microbial and molecular genetics, large quantities of bacteria and highly purified phage stocks are necessary. Generally, culturing of bacteria causes no methodological difficulties, provided that suitable culture vessels and a centrifuge of an adequate capacity are at hand. The growth, concentration and purification of phages, however, is more complex. Therefore in the following three experiments we want to demonstrate these methods for phage T4. T4 serves as an example and is representative for many other phages.

The widespread use of analytical and preparative ultracentrifuges and their importance in protein and nucleic acid research convinced us to show the two centrifugation techniques most frequently used in molecular biology. In experiment No. 4 two phages of different density will be separated by sedimentation equilibrium centrifugation in a CsCl density gradient. In experiment No. 5 proteins of different molecular weights will be separated on a sucrose gradient and the sedimentation constant s of one of the proteins will be estimated by comparison with the sedimentation constants of two reference proteins.

Literature

BOWEN, T.J.: An Introduction to Ultracentrifugation. London: Wiley Interscience 1971.

CANTONI, G.L., DAVIES, D.R. (eds.): Procedures in Nucleic Acid Research. London: Harper & Row 1967.

DIRKX, S.J.: Analytical Density Gradient Centrifugation. Pamphlets of Beckman Instruments Inc.

FRAENKEL-CONRAT, H., WAGNER, R.R. (eds.): Comprehensive Virology, 5 Volumes. New York: Plenum 1974-1976.

HABEL, K., SALZMANN, N.P.: (eds.): Fundamental Techniques in Virology. London: Academic Press 1969.

MARAMOROSCH, K., KOPROWSKI, H. (eds.): Methods in Virology, 4 Volumes. London: Academic Press 1967-1968.

NORRIS, J.R., RIBBONS, D.W. (eds.): Methods in Microbiology, 8 Volumes now in print. London: Academic Press 1969-1973.

PERRY, E.S., van OSS, C.J. (eds.): Progress in Separation and Purification. London: Wiley Interscience 1969.

1. Large Scale Production of Phage T4

Phages are viruses which can only infect bacteria. The sequence
of events in this process is as follows: Adsorption of the phage
to specific receptors on the bacterial cell wall, injection of
the phage nucleic acid into the host cell, transcription and
replication of this nucleic acid, synthesis of proteins, specific aggregation of viral subunits (maturation) and finally,
lysis of the host cell and release of the phage progeny. Under
optimal growth conditions only 24 min elapse between adsorption
of the T4-phages to the bacteria and the lysis of the host
("latent period"). The average burst size for T4 is about 100
phages per infected cell.

T4 is a virulent phage. Electron micrographs reveal a complex
structure endowed with a head and a tail, both of approximately
the same length (total length = 200 nm). The head and the tail
are each composed of several different proteins and exhibit
substructures. The DNA which is tightly packed in the phage
head is a linear double-stranded molecule. Its molecular weight
is about 130×10^6 daltons. It contains approximately 65 mole
percent (A + T). Cytosine is replaced by 5-hydroxymethylcytosine
(HMC). In addition, one molecule of glucose is attached to the
hydroxymethyl group of each HMC residue. T4 is one of the best
investigated phages.

The large-scale production of virulent phages described in this
experiment will often be modified under research conditions,
e.g. minimal medium might be used instead of nutrient broth.
Stocks of temperate phages are frequently obtained in another
manner. A short outline will be given at the end of this experiment.

Plan. A growing culture of *E. coli* will be infected with phage
T4. The growth of the bacteria and, at a later time, the phage-
induced lysis will be followed microscopically (cell count) and
photometrically (cell mass). After several hours of incubation,
a crude lysate free of most of the bacterial debris will be
obtained by centrifugation.

Material. One 2000-ml Erlenmeyer flask containing 1000 ml
HERSHEY broth and an aeration device. The flask stands in a
37°C water bath. Air pressure connection. 30 ml stat culture
of *E. coli* BA in NB (cell titer 2×10^9/ml). 3 ml log culture
of *E. coli* BA as an indicator (cell titer 5×10^8/ml). 1 ml
phage suspension with a titer of 2×10^{10}/ml: for even-numbered
student groups, the osmotic shock-resistant mutant T4o; for odd-
numbered student groups, a T4rII rapid lysis mutant. 4 plates
with NB-agar. 4 tubes with each 3 ml of soft agar. 6 centrifuge
bottles (250 ml each). 1 ml antifoam. 1 graduated cylinder
(200 ml) and one 1000 ml flask. 1 sheet of semi-logarithmic
paper. For several groups together, 1 SORVALL-centrifuge
RC2-B with GSA rotor and 6 centrifuge bottles.

Procedure

1. __Bacterial culture.__ Inoculate the HERSHEY broth with 30 ml of the stat culture of _E. coli_ BA and start aeration. Immediately (t = 0) take a 10 ml sample to measure the optical density (O. D.) at 580 nm in a photometer. Use the same cuvette for all further readings (to be taken at 20 min intervals). Use a second cuvette filled with sterile HERSHEY-broth as reference. If round cuvettes are used, they must be marked so that they can always be read in the same position in the photometer. At the same time as the O. D. readings are taken, the cells are counted in a counting chamber on a phase contrast microscope (magn. 10 × 40). In order to obtain the cell count per ml ("cell titer, N"), multiply the average number of cells per small square ($\Sigma n/\Sigma i$) of the counting chamber, with the chamber factor (often 2×10^7). After the measurement return the 10 ml samples to the Erlenmeyer flask. Record the measured cell titer and the O. D. readings in a table. Graph logarithms of these values as a function of the incubation time (t). If n ⩾ 10 dilute the samples 1:5 or 1:10 in P-buffer accordingly, but do not return these samples into the Erlenmeyer flask.

2. __Phage infection.__ After an incubation time of about 90 min, the culture reaches a cell density of 4×10^8/ml. Add the 2×10^{10} phages to the culture, mix immediately. Continue incubation and aeration as well as O. D. readings and cell counts. Formation of foam must be prevented by addition of a drop of sterile antifoam-emulsion. 4 hrs after addition of the phage, stop the aeration, add approx. 5 ml chloroform, swirl and keep the culture first in a 37°C water bath and then in an ice-bath for about 30 min each.

3. __Phage harvest.__ Fill 175 ml into each centrifuge bottle (use graduated cylinders). Centrifuge the tightly closed bottles in a GSA rotor of a SORVALL centrifuge RC2-B for 7 min at 7500 rpm. Decant the supernatant ("crude lysate") into a 1000-ml flask and discard the sediment.

4. __Phage-titer.__ Dilute a 0.1 ml sample of the crude lysate 10^{-7} and 10^{-8} in P buffer. Pipette 0.1 ml of the corresponding dilution and 0.2 ml of _E. coli_ BA (log culture) as indicator, into tubes with 3 ml soft agar and plate

Dil.	Plate No.	Plaques	Titer
10^{-7}	1		
10^{-7}	2		
10^{-8}	3		
10^{-8}	4		

Incubate plates at 37°C.

5. <u>Evaluation.</u> The next day count the plaques on plates No. 1-4 and calculate the phage titer of the crude lysate. Calculate from the exponential part of the growth curve (log N versus t):

Growth constant $\qquad$ $k = 2.30 \ (\log N - \log N_0)/t \ (\mathrm{min}^{-1})$

Number of generations $\quad$ $g = k \times t/\ln 2$

Generation time $\qquad$ $T = t/g \ (\mathrm{min})$

$\ln 2 = 0.69$

For further details see Appendix.

<u>Preparation of high titer stocks of temperate phages</u> requires induction of the culture lysogenic for this phage by UV-irradiation or, if the prophage has a temperature-sensitive repressor, a short heat treatment followed by incubation until lysis. Another technique is the following:

Two drops of an overnight culture of a phage-sensitive bacterial strain and 10^5 to 10^6 phage particles are mixed in 3 ml of NB soft agar and are layered onto a NB agar plate. After incubation of several such plates for 6-18 hrs the agar layers are scraped off with a glass rod and are pooled in a flask. 3-5 ml of nutrient broth or buffer are added per plate and the mixture is vigorously shaken or stirred in order to facilitate elution of the phage particles from the soft agar. Subsequently agar and bacterial debris are removed by low speed centrifugation. For further details see ADAMS, p. 456.

<u>Literature</u>

ADAMS, M.H.: Bacteriophages (esp. p. 454-460). London: Interscience Publ. Inc. 1959.

EISENSTARK, A.: Bacteriophage Techniques. In: Methods in Virology 1, 450-524. London: Academic Press 1967.

SARGEANT, K.: Large-Scale Bacteriophage Production. In: Advances in Applied Microbiology 13, 121-137 (1970).

STENT, G.S.: Molecular Biology of Bacterial Viruses. London: Freeman & Comp. 1963.

THOMAS, C.A., Jr., ABELSON, J.: The Isolation and Characterization of DNA from Bacteriophage. In: Procedures in Nucleic Acid Research (eds. G.L. CANTONI, D.R. DAVIES) pp. 553-555. London: Harper & Row 1966.

The first two titles are general textbooks on bacteriophages whereas the rest deals specifically with phage enrichment and purification-techniques.

<u>Time requirement:</u> 1st day 7.5 hrs, 2nd day 0.5 hrs.

Data sheet

Duration (t) of Experiment hrs min	Time	O.D.$_{580}$	$\frac{\Sigma n}{\Sigma i}$[a]	Cell titer (N)	Remarks[b]
0 0 20 40					
1 0 20 40					
2 0 20 40					
3 0 20 40					
4 0 20 40					
5 0 20 40					
6 0 20 40					

[a] $\Sigma n/\Sigma i$ = Number of cells counted per number of the small squares observed.

[b] Remarks: e.g. addition of phage, antifoam or chloroform, or: diluted 1 : 5 or 1 : 10 for cell counts.

Appendix to 1: Growth Curve of Bacteria

"Growth" specifies the irreversible increase of the cell mass. The growth of bacteria expresses itself first as cell enlargement and then as cell division. The term "growth curve" means the semi-logarithmic plot of the cell titer, viable count or cell mass in a liquid culture as a function of incubation time.

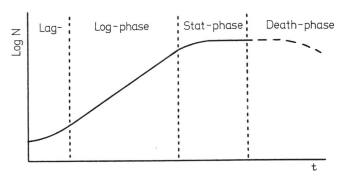

Cell titer: Cells per ml of culture, i.e. living and dead cells, mostly determined by microscopic counts in a suitable counting chamber. Germ titer (viable count): Cells per ml of culture which have the ability to divide ("germing"), i.e. those which can form colonies if plated on nutrient agar. Cell mass: The amount of cellular substance per ml of culture, e.g. to be determined either photometrically (O.D.$_{580}$), or as wet or dry weight. A stat phase culture and a log phase culture may have the same cell titer, but may differ in cell mass by a factor of 3.

The growth curve can be divided into: (lag-) phase: The period of time, beginning with inoculation of the culture during which the cells begin to divide, and ending when the division rate reaches its maximum. Exponential (log-) phase: The period of time in which the maximum rate of division is maintained. The average time for the doubling of cell mass, cell titer and viable count is constant. Stationary (stat-) phase: The period of time during which the increase of cell mass and of cell titer and viable count slows down and finally comes to a stop. Death phase: The period of time during which the viable count and later the cell titer and cell mass decrease as a result of cell death and autolysis.

N = cell titer or viable count

t = time of incubation (min)

k = growth constant (min^{-1})

g = generations, i.e. number of cell divisions

T = average generation time (min)

In the exponential growth phase, each cell yields two daughter cells in regular intervals (2^0, 2^1, 2^2 ...2^g). The increase in cell number from N_0 to N in the physical time unit, t, is speci-fied by the differential equation:

$$\frac{dN}{dt} = k \times N \text{ or } k \times dt = \frac{dN}{N}. \tag{1}$$

Integrated this equation writes as

$$k \times t = \ln N + C \tag{2}$$

where C is a constant. At the beginning of the log phase, that is at t = 0, N equals N_0. Therefore one can write

$$k \times 0 = \ln N_0 + C \text{ or } C = -\ln N_0 \tag{3}$$

and by insertion into the Eq. (2) one gets:

$$k \times t = \ln N - \ln N_0 \quad \text{or} \tag{4a}$$

$$\underline{k = 2.30 (\log N - \log N_0)/t \ (\min^{-1})} \tag{4b}$$

because $\ln N = 2.30 \times \log N$.

Eq. (4a) written in the following form

$$\ln N = \ln N_0 + k \times t \tag{4c}$$

is synonomous with the standard equation of the first order $y = b + ax$, i.e. <u>k gives the slope</u> of the growth curve. Eq.(4c) can also be written as an exponential equation

$$N = N_0 \times e^{kt}. \tag{4d}$$

If the physical time unit, t, is replaced by a biological one, for instance the number of generations, g (where T = time for 1 cell division) cell growth can also be described by the fol-lowing equation

$$N = N_0 \times 2^g. \tag{5}$$

Equating (4d) and (5):

$$N_0 \times e^{kt} = N_0 \times 2^g \quad \text{or} \quad k \times t = g \times \ln 2 \tag{6a}$$

or the number of generations $\underline{g = k \times t/\ln 2}$ \hfill (6b)

$$\ln 2 = 0.69.$$

The average generation time, T, is given as

$$\underline{T = t/g \ (\min)} \tag{7}$$

Spectrophotometric Measurement of Cell Density

When a <u>solution</u> is brought into the light path of a spectropho-meter and the light intensity before (I_0) and after (I) passing

through the solution is measured, the concentration (c) of the
absorbing substance is proportional to the extinction E (synonymous to optical density and absorption).

$$E = \log \frac{I_0}{I} = \varepsilon \times l \times c \tag{8}$$

l = length of light path in the sample (cm)

ε = molar extinction coefficient (cm^2/ mmole)

c = molar concentration (mole/l)

If a growth curve is followed photometrically, e.g. at a wave
length of 580 nm, N can be replaced by E and (4b) reads as
follows:

$$k = 2.30 \ (\log E - \log E_0)/t \ (\min^{-1}). \tag{9}$$

The optical density at λ = 580 nm obtained with a suspension
of non-pigmented bacteria is exclusively due to light scattering
by the cells. In a certain range the number of cells per ml is
proportional to the optical density of the suspension. Measurements of light scattering by bacteria is also possible at other
wavelengths in the range 400 - 700 nm.

2. Purification and Concentration of Phage T4

a) Dextransulfate – PEG Two-Phase Technique

A crude lysate contains the phages and cellular debris plus dissolved components of the lysed host bacteria. The impurities can be easily removed within one day by the sodium dextransulfate-polyethylene glycol two-phase technique. Furthermore, this method also concentrates the phages.

Sodium-dextransulfate "500" is a water-soluble polyelectrolyte; its average molecular weight is 500,000.

Polyethylene glycol (PEG), "CARBOWAX 6,000", is another water-soluble polymer with a molecular weight of 6,000 - 7,500. If aqueous solutions of these two substances are mixed, within a short time they will separate into two phases. If still other dissolved or suspended substances are present, e.g. proteins or phages, these will be distributed in a characteristic way in both phases. Their distribution depends on the ionic strength (0.1 - 1 M NaCl) and on the concentration of PEG. It is characterized by the "partition coefficient, K": $K = (C_u)/(C_l)$, i.e. the concentration in the upper phase (C_u) divided by the concentration in the lower phase (C_l). Equal distribution in both phases would therefore mean $K = 1$. The dextransulfate solution always forms the lower phase. Under suitable conditions, the phages concentrate at the interphase and the bacterial debris move into the upper phase.

Plan. In order to find the optimal conditions for purifying and concentrating phage T4 by the two-phase technique a series of tests will be performed as follows.

Constant amounts of T4 crude lysate (2 ml) and dextransulfate solution (0.4 ml) will be mixed with various amounts of concentrated PEG-solution. The disappearance of the phages from the upper phase is to be measured. Proof that the phages have really been removed from the upper phase and were not simply inactivated will not be given here (see Appendix).

Material. 20 ml T4 crude lysate, e.g. from Expt. 1, diluted in NB to a titer of 5×10^8/ml. 5 ml 20% (w/v) aqueous Na-dextransulfate "500" solution (PHARMACIA). 10 ml of a 30% (w/v) aqueous polyethylene glycol, "Carbowax 6,000", solution (Union Carbide). One 100 ml flask with 0.48 g NaCl. 18 plates with NB-agar. 18 NB-soft agar tubes. 5 ml log culture of *E. coli* BA grown in NB, as an indicator (approx. 5×10^8/ml). One sheet of semi-logarithmic paper. For several groups together: 1 super mixer (e.g. Vortex). 1 centrifuge, suited for low speed centrifugation of small test tubes.

Procedure

1. <u>Start.</u> Dissolve 0.48 g NaCl in 20 ml T4 crude lysate in the sterile flask. Pipette into 6 small sterile test tubes marked 0 - 0.3 - 0.6 - 0.9 - 1.2 and 1.5:

   2.0 ml NaCl-containing lysate

   0.4 ml Na-dextran sulfate solution

   0 to 1.5 ml PEG-solution respectively (see data sheet)

Mix the samples for a short time on the Vortex mixer and allow to stand at 4°C for 30 min.

2. <u>Centrifugation and titering.</u> Centrifuge the 6 test tubes at low speed for 10 min. This will hasten the separation of both phases.

Carefully remove 0.5 ml from the upper part of each of the 6 upper phases, dilute each sample 10^{-3}, 10^{-4} and 10^{-5} in P-buffer, and plate, according to the agar layer technique, 0.1 ml samples of each with 0.2 ml *E. coli* BA as indicator, on NB-agar plates (Nos. 1-18). Incubate plates overnight at 30°C.

3. <u>Evaluation.</u> Count the plaques on plates 1-18 and calculate the titers. Multiply each phage titer by a correction factor (see data sheet). Since variable amounts of PEG-solution were given to each 2 ml of the crude lysate, it is necessary to do this in order to make the results comparable. Plot the corrected titers semi-logarithmically as a function of the amount of PEG. For this purpose express all titers as a fraction (%) of the titer of the control marked zero (0 ml PEG).

- What was the minimum amount of PEG that removed most of the phages from the upper phase? Estimate the corresponding partition coefficient assuming that the phages which were removed from the upper phase are now in the lower phase.

- What is the final molar concentration of the NaCl (MW = 58.5) in the test tube with 0.9 ml PEG-solution?

- PEG could have been added to the NaCl-containing crude lysate in dry form in order to avoid dilution. How many grams of PEG would have to be dissolved in 10 l crude lysate, if as many phages as possible were to be collected in the interphase?

Appendix

The test could be continued in the following way: Completely remove the upper and lower phases of those test tubes, having very little phage left in the upper phase. Use PASTEUR pipettes. Dissolve the remaining interphase in as little distilled water (or 1% sodium dextransulfate) as possible, pool and add 3 M KCl-solution (0.15 ml KCl per 1.0 ml dissolved interphase). After

standing for 2 hrs, the sodium dextransulfate which was precipitated with KCl is removed by low speed centrifugation for
10 min. The supernatant contains highly concentrated and purified phages. This step can be repeated until no further precipitate is formed by KCl.

Literature

ALBERTSSON, P.-Å.: Two-Phase Separation of Viruses. In: Methods
in Virology 2, 3o3-321 (1967).

ALBERTSSON, P.-Å.: Partition of Cell Particles and Macromolecules in Polymer Two-Phase Systems. In: Advances in
Protein Chemistry 24, 303-341 (1970).

Time requirement: 1st day 2 hrs, 2nd day 2 hrs.

Data sheet

ml PEG[a]	Dilution	Plate No.	Plaques	Phage titer uncorr. corr.		F[b]
0	10^{-3} 10^{-4} 10^{-5}	1 2 3				1.00
0.3	10^{-3} 10^{-4} 10^{-5}	4 5 6				1.15
0.6	10^{-3} 10^{-4} 10^{-5}	7 8 9				1.30
0.9	10^{-3} 10^{-4} 10^{-5}	10 11 12				1.45
1.2	10^{-3} 10^{-4} 10^{-5}	13 14 15				1.60
1.5	10^{-3} 10^{-4} 10^{-5}	16 17 18				1.75

[a] Identical with the mark on the test tubes.

[b] F = correction factor, e.g. F = (2.0 + 1.2)/2.0 = 1.60.

3. Purification and Concentration of Phage T4

b) Centrifugation with and without CsCl

If a cell-free crude lysate of phage T4 is centrifuged at high speed (20,000 × g) for 1-2 hrs, the phages will sediment together with the debris of the host cells. By resuspending the sediment in a volume smaller than the original one, concentrations of up to 50-100 fold can be easily achieved. If such a resuspension is layered on a CsCl solution and is centrifuged at high speed the phages will sediment into the CsCl solution and form a compact band in 3-4 hrs. The cell fragments and dissolved impurities, because of their lower density, will remain in the upper layer. The region of the CsCl solution which contains the phage is dialyzed to remove the CsCl. The product is a highly purified and highly concentrated phage suspension.

Plan. A crude lysate of T4-phages is to be sedimented by centrifugation, resuspended in a small volume of P-buffer and recentrifuged at high speed. For the second centrifuge run the phage suspension is to be layered on a CsCl solution of high density. The phages which sediment into the CsCl phase form a sharp opalescent band. Fractions are collected by "dripping" the contents of the centrifuge tube through a hole punched into the bottom of the tube. The fractions containing the phage will be pooled, dialyzed and titered. The absorption spectrum of the purified phage suspension will be compared with that of a solution of T4-DNA.

Material. About 900 ml of a T4o lysate from Expt. 1 with a titer of $10^{10}-10^{11}$ phages/ml. 5 ml of a log culture of *E. coli* BA in NB (cell titer about 5 × 10^8/ml). 4 NB-agar plates. 4 NB-soft agar tubes. One 200-ml graduated cylinder. 6 250-ml centrifuge bottles for the GSA-Rotor and two 10-ml tubes for the SS34 Rotor of the SORVALL RC2-B centrifuge. 2 cellulose nitrate centrifuge tubes No. 853 for the SW25.1 rotor of a BECKMAN-ultracentrifuge. Two PASTEUR pipettes with rubber bulbs. One pin. About 10 cm of dialysis tubing, 1,000 ml of P-buffer containing 0.002 M Mg^{++}. One magnetic stirrer. One sheet of linear graph paper. About 15 ml of an aqueous solution of CsCl, density = 1.55 g/cc*. For several groups together: One SORVALL RC2-B centrifuge with GSA and SS34 rotors. One BECKMAN ultracentrifuge (Spinco L50) with an SW 25.1 swinging bucket rotor. One ZEISS refractometer, with thermostat, kept at 25°C. One balance. 3 200-ml beakers (for taring the filled swinging buckets). One UV-spectrophotometer with 3 quartz cuvettes (1 cm light path). Paraffin, liquid. 3 ml of a DNA solution (about 25 µg/ml) in P-buffer, (preferably T4-DNA).

---

* CsCl-solutions are prepared according to the following formula: g CsCl per 100 cc aqueous solution = 137.48 - 138.11 (1/ρ). The density at 25°C is measured by the refractive index of the CsCl solution (see standard curve, p. 42).

Procedure

General: Use only phage T4o for purification by CsCl centrifugation; other T4 phages could be osmotically shocked during dialysis.
Pool high titer T4o lysates from several groups and record the
volume of the lysates mixed together as well as their titers.

1. **Pre-purification.** Fill 175 ml of lysate into each centrifuge
bottle and centrifuge at 11,000 rpm (19,600 × g) for 90 min in
the GSA rotor (for smaller volumes use the SS34 rotor and centrifuge at 18,000 rpm (39,100 × g) for 45 min). Discard supernatants. Immediately resuspend each pellet in 8 ml or even less
of P-buffer by blowing the buffer with a pipette repeatedly over
the pellet until it is resuspended. If more time is available,
phage pellets may be overlayered with buffer and allowed to sit
overnight to resuspend. Centrifuge the resuspension 10 min at
low speed (SS34 rotor, approx. 4,500 rpm) and use only the
supernatant.

2. **Purification with CsCl.** Fill about 24 ml of phage suspension
obtained in step 1 into centrifuge tubes (No. 853) fitting the
SW25.1 rotor of a BECKMAN-ultracentrifuge, and carefully layer
under each phage suspension 6 ml of the CsCl-solution (density
1.55 g/cc). Use a PASTEUR-pipette. Then balance the tubes by
addition of ∿1 ml of paraffin. Check weight on a balance. After
3 hrs of centrifugation at 20,000 rpm (approx. 40,000 × g),
pierce the bottom of each centrifuge tube with a pin and drip
only the opalescent phage band into small sterile test tubes
(1-2 ml per tube).

3. **Dialysis.** Dialyze first against 1,000 ml plain P-buffer for
2 hrs, and afterwards overnight against 1,000 ml P-buffer containing Mg^{++}. Stir magnetically. Mg^{++} is omitted from the first
buffer in order to avoid precipitation of the phage in the dialysis bag in the presence of a high concentration of CsCl.

4. **Spectrophotometry** (see data sheet). Dilute a sample of the
dialzyed phage suspension about 10^{-2} in P-buffer. Read the absorption spectrum between 220-320 nm in intervals of 5 nm. P-
buffer serves as a blank and the DNA solution (25µg/ml) as a
control. The optical density (O.D.) of the phage suspension is
not only due to the absorption of light by the phage protein and
nucleic acids but also to the scattering of light by the phage
particles. At 320 nm the O.D. of the phage suspension is entirely due to light scattering. The real absorption can be calculated
according to the formula of RAYLEIGH. All O.D. values measured
between 220 and 315 nm (O.D. $_{uncorr.}$) have to be corrected in the
following way:

$$\text{O.D.}_{corr.} = \text{O.D.}_{uncorr.} - (\frac{320}{\lambda})^4 \times \text{O.D.}_{320}.$$

All O.D. values (uncorr.; corr.; DNA) will be plotted graphically
as a function of the wave length.

5. _Titering._ Dilute the dialyzed phage suspension 10^{-8} and 10^{-9} in P-buffer. Plate 0.1 ml samples with 0.2 ml of _E. coli_ BA as an indicator by the soft agar overlay technique on nutrient agar.

Dilution	Plate No.	Plaques	Titer
10^{-8}	1		
10^{-8}	2		
10^{-9}	3		
10^{-9}	4		

After 18-hrs incubation at 37°C plaques will be counted and the titers calculated.

Evaluation

1. _Degree of concentration and phage yield._ Compare titer A of the dialyzed phage suspension with titer B of the mixture of crude lysates:

$$\frac{A}{B} = \frac{\rule{3cm}{0.4pt}}{} = \underline{\underline{\rule{3cm}{0.4pt}}} \quad .$$

Compare Volume C of the mixture of crude lysates used with Volume D of the dialyzed phage suspension:

$$\frac{C}{D} = \frac{ml}{ml} = \underline{\underline{\rule{3cm}{0.4pt}}} \quad .$$

If the ratios differ, discuss possible reasons.

2. _Purity_

Ratio of O.D. Values	Suspension dial.phages (corrected values)	DNA solution
260/280		
260/235		

Why are the ratios for the DNA solution larger than those for the phage suspension?

$$\frac{O.D._{260}; \text{corr.}}{1 \times 10^{12} \text{ phages/ml}} = \underline{\underline{\rule{3cm}{0.4pt}}} \quad .$$

Literature

ANDERSON, N.G., CLINE, G.B.: New Centrifugal Methods for Virus
  Isolation. In: Methods in Virology 1, pp. 137-178. London:
  Academic Press 1967.

EICHENBERGER, W.: Trennung biologischer Partikel durch Zentri-
  fugation im Dichtegradienten. Chimia 23, 85-94 (1969).

VINOGRAD, J., HEARST, J.E.: Equilibrium Sedimentation of Macromolecules and Viruses in a Density Gradient. Fortschritte der
  Chemie Organischer Naturstoffe 20, 372-422 (1962).

Time requirement. 1st day 7-8 hrs, 2nd day 2 hrs, 3rd day 0.5 hrs.

Data sheet

Wave length λ (nm)	Optical density (O.D.) Phages uncorr.[a]	phages corr.	O.D. DNA soln.	$(\frac{320}{\lambda})^4$	$(\frac{320}{\lambda})^4 \times O.D._{320}$
225				4.08	
230				3.74	
235				3.43	
240				3.16	
245				2.91	
250				2.68	
255				2.47	
260				2.29	
265				2.12	
270				1.97	
275				1.83	
280				1.70	
285				1.59	
290				1.48	
295				1.38	
300				1.29	
305				1.21	
310				1.13	
315				1.06	
320				1.00	

[a] Suspension of phages, after CsCl density gradient centrifugation and dialysis. The plaque titer is ...../ml.

4. Determination of the Density of Phages by CsCl Gradient Centrifugation

If concentrated, aqueous solutions of CsCl or Cs_2SO_4 are centrifuged at high speed for sufficient time, a stable and approximately linear density gradient builds up in the gravitational field. This "isopycnic gradient" is formed because a sedimentation-diffusion-equilibrium is established which depends on the centrifugal force and the concentration of the salt solution. The densest fractions correspond to the highest concentrations of salt and are found on the bottom of the centrifuge tube. If viruses or macromolecules are added to the salt solution, these band in the particular zone of the gradient which corresponds to their own density. Substances with different buoyant densities can be concentrated in different zones of the centrifuge tube by this procedure, and can be separated from one another by fractionating the gradient.

In the following experiment, the wild-type and the deletion mutant cb_2b_5* of the temperate phage λ will be separated. Both phages have equal size and shape. The wild type phage is composed of 50% protein (density ≈ 1.3 g/cc) and 50% double-stranded DNA (density ≈ 1.7 g/cc); therefore, its total density approaches 1.5 g/cc. If the phage loses part of its chromosome by deletion or by the formation of a transducing particle (λdgal), its density decreases. This decrease in density can be determined in the CsCl-gradient. Many of the basic concepts of molecular genetics, e.g., semiconservative replication and strand-selective transcription, were arrived at by experiments made possible by the finding that nucleic acids can be labeled with density markers and can then be quantitatively separated by CsCl-gradient centrifugation. To introduce a density marker, ^{15}N is frequently substituted for ^{14}N, Br for CH_3 and ^{13}C for ^{12}C.

Plan. A mixture of a wild-type λ and the deletion mutant cb_2b_5 will be separated by CsCl-gradient centrifugation. The gradient will be fractionated and its slope determined by measuring the refractive index of selected fractions. The position of the phage bands in the gradient will be determined by a rapid and semi-quantitative drop test. The density of both phage types and the relative length of the DNA-segment lost by deletion will be estimated.

Material. 1st day: 0.4 ml of a suspension of phages λ wild type and λcb_2b_5 resp. (1×10^8/ml). Crystalline CsCl. Paraffin, liquid. 1 polyallomer centrifuge tube (BECKMAN No. 249). 3 PASTEUR pipettes. One 100-ml beaker. For several groups

---

* In this mutant the deletions b_2 and b_5, which arose independently from one another, as well as the point mutation c (clear plaque), were introduced into the same genome by recombination.

together: 1 scale. 1 refractometer with thermostat. 1 BECKMAN
ultracentrifuge with swinging bucket rotor SW56Ti.

2nd day: 1 ml of a log culture of *E. coli* K12s in NB medium (cell
titer approx. 8×10^8/ml). 5 plates with TBY-agar. 5 TBY-soft
agar tubes. Test tube rack with 30 small test tubes. 80 ml
nutrient broth. 1 sheet of graph-paper. For several groups together: 1 gradient dripping device with hypodermic needle of
0.6 mm inner diameter (see p. 41), or simply a pin to puncture
the tube.

Procedure

1. Preparation of CsCl solution: To prepare a CsCl solution of
a given density, the following formula is applied:

Percent of weight of CsCl in solution = $137.48 - (138.11 \frac{1}{\rho_{25^\circ}})$

where ρ = density in grams per cc at 25°C. In our experiment,
$\rho_{25^\circ} = 1.4790$ g/cc, so that the wild type and the deletion mutant both band approximately in the middle of the gradient. According to the above equation, 44.10 g of CsCl must be contained
in 100 g of aqueous solution. Since only 10 g of CsCl solution
per group are needed, only 1/10 of this amount will have to be
prepared:

> 4.410 g CsCl
>
> 5.090 g distilled water
>
> 0.250 ml λ wild type $(1 \times 10^8$/ml)
>
> 0.250 ml λ cb$_2$b$_5$ $(1 \times 10^8$/ml)

Sum: 10.000 g

Check the density of the prepared CsCl solution in the refractometer at 25°C. Read the relationship between density and refractive index from the diagram (p. 42). A density of $\rho = 1.4790$
g/cc corresponds to a refractive index of $\eta = 1.3790$. If the refractive index measured is higher or lower, it can be corrected
by the careful addition of distilled water or crystalline CsCl.

2. Centrifugation: Pipette 3.5 ml of CsCl solution containing
the phages into a polyallomer tube, overlayer with paraffin,
tare against tubes from other groups and centrifuge in the
SW56Ti rotor at 23,000 rpm (55,000 $\times$ g) for approx. 24 hrs at
20°C.

3. Dripping the gradient (2nd day): Pierce the tubes with the
dripping device and collect fractions of 12 drops each in small
test tubes. Measure the refractive index of every 5th fraction
at 25°C beginning with the last fraction and plot against the
fraction number. Dilute the rest of the fractions by the addition of 2.5 ml NB to each. Mark 6 sectors on the bottom of each
of five petri dishes and number the sectors from 1-30. Layer

these plates with a soft agar layer containing 0.1 ml of *E. coli*
K12s per plate. After hardening of the soft agar layer, spot
1 drop of each fraction into the corresponding sector, using a
sterile inoculating loop. Do not turn the plates upside down
and do not tilt, so that the drops do not run. Incubate the
plates overnight at 37°C.

Data sheet

Fraction No.	Refractive index η	Plaques per sector	Fraction No.	Refractive index η	Plaques per sector
1			16		
2			17		
3			18		
4			19		
5			20		
6			21		
7			22		
8			23		
9			24		
10			25		
11			26		
12			27		
13			28		
14			29		
15			30		

Evaluation (3rd day)

1. Count or, if counting is no longer possible, estimate (+, ++,
+++), the number of plaques in the sectors. Record the results
and plot against the fraction number. Compare the curves and
determine the density of both phages from the diagram on p. 42.

Phage	Refractive index η	Density ρ
Wild type		
cb_2b_5		

2. Calculate the percent of loss, α, of DNA in the deletion mutant as opposed to the DNA content of the wild-type phage according to WEIGLE et al. (1959):

$$\alpha = \frac{2 \times \Delta\rho}{0.21 - \Delta\rho} = \underline{\qquad\qquad} .$$

Literature

BRAKKE, M.K.: Density-Gradient Centrifugation. In: Methods in Virology 1, pp. 93-118. London: Academic Press 1967.

KELLENBERGER, G., ZICHICHI, M.L., WEIGLE, J.: A Mutation Affecting the DNA Content of Bacteriophage Lambda and its Lysogenizing Properties. J. Biol. Chem. 3, 399-408 (1961).

SZYBALSKI, W.: Use of Cesium Sulfate for Equilibrium Density Gradient Centrifugation. In: Nucleic Acids, Part 2, pp. 330-360. London: Academic Press 1968.

WEIGLE, J., MESELSOHN, M., PAIGEN, K.: Density Alterations Associated with Transducing Ability in the Bacteriophage Lambda. J. Mol. Biol. 1, 379-386 (1959).

Time requirement: 1st day 2-3 hrs, 2nd day 3 hrs, 3rd day 1 hr.

Gradient dripper

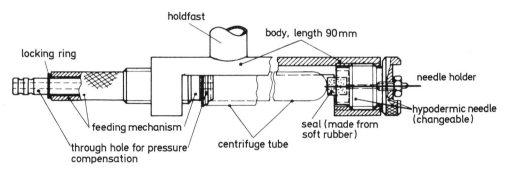

Material: Brass

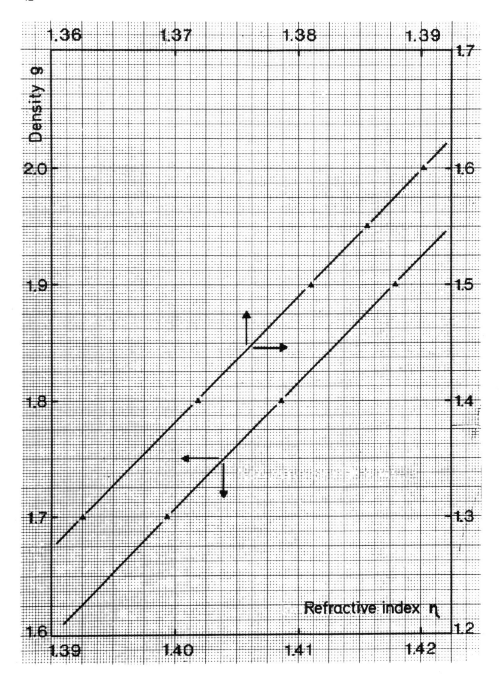

Refractive index η_{25° of an aqueous solution of CsCl as a function of the density ρ_{25°. (The arrows indicate the coordinates for that line)

5. Estimation of Sedimentation Constants by Sucrose Gradient Centrifugation

The sedimentation constant s, also called the s-value, is part of the SVEDBERG equation for the calculation of the molecular weights of macromolecules. It can be determined by analytical ultracentrifugation.

$$s = \frac{dx/dt}{\omega^2 x} . \qquad (1)$$

dx/dt = the speed with which the substance sediments in the gravitational field
ω = the angular speed
x = distance of the substance from the axis of rotation.

The unit in which sedimentation constants are given is S (for SVEDBERG).

$$1 \ S = 10^{-13} \ sec.$$

The rate of sedimentation of a macromolecule in a gravitational field depends on its molecular weight as well as on its second- ary and tertiary structure. For example, the sedimentation con- stant of single-stranded DNA generally is larger than that of double-stranded DNA of the same molecular weight. Transfer RNA of different origin has an s-value of 4.5 S, native DNA of phage λ has a value of 33.6 S and that of phage T4 61.3 S. For most proteins, the s-values lie somewhere between 1 and 50 S.

The simplest method of estimating the s-value of a substance is to compare its sedimentation speed with that of another (refer- ence) substance having a known s-value; for this comparison, how- ever, only substances of about the same partial specific volume as that of the substance under investigation are suitable. Thus, to estimate the s-value of a spherical protein, only spherical proteins should be taken as a reference. For double-stranded DNA only double-stranded DNA standards should be used etc. Both sam- ples are layered on a sucrose gradient in a thin band. The smaller the band of the layered sample material, the better the separa- tion achieved. Because of their differing sedimentation speeds, both substances separate in the gravitational field.

The s-values of the sample and of the standard are then related by the equation

$$s_{20w} \ unknown = s_{20w} \ reference \times (D_1/D_2) \qquad (2)$$

$$D_1/D_2 = \frac{Distance \ travelled \ by \ the \ unknown \ substance}{Distance \ travelled \ by \ the \ reference \ substance}.$$

In this case the distance is measured from the meniscus and recorded in "number of fractions". s_{20w} is the corrected sedimentation constant, calculated from the viscosity and density of water at 20°C (standard conditions).

The linear gradient of sucrose concentration, usually between 5 and 20%, diminishes convection currents. It also keeps the speed of the sedimenting molecule approximately constant since it counteracts the molecule's acceleration in the gravitational field. The acceleration of the macromolecule increases with increasing distance from the axis of rotation and is balanced by the increasing viscosity.

In contrast to the CsCl-gradient (Expt. 4), which forms in the gravitational field, sucrose gradients must be pre-formed. This is done by the use of a simple gradient mixer or alternating by careful layering of sucrose solutions of linearly decreasing concentration. Finally, the gradients are allowed to stand for 15 hrs at 4°C in order to allow discontinuities to smooth out by diffusion. Instead of nucleic acids we will use various, well-defined and readily available proteins in the following exercise, in order to simplify the preparation and evaluation.

<u>Plan.</u> Three globular proteins, cytochrome c, lactate dehydrogenase and β-galactosidase will be centrifuged together on the sucrose gradient. Cytochrome c (1.7 S) and β-galactosidase (15.9 S) serve as reference molecules. The s-value of lactate dehydrogenase will be determined.

<u>Material. 1st day:</u> 5 ml each of 5% and 20% sucrose solution (W/V) in 0.1 M phosphate buffer, pH 7, containing 0.1 M β-mercaptoethanol. 1 polyallomer centrifuge tube (BECKMAN No. 249). 1 PASTEUR pipette. For several groups together: 1 gradient mixer (volume 10 ml) with stirring motor. Slices of cork. 1 pair of forceps. 1 BECKMAN ultracentrifuge with swinging bucket rotor SW56Ti. Paraffin, liquid. 2 ml sample mixture, containing 4 mg/ml cytochrome c, 20 μg/ml lactate dehydrogenase (beef heart) and 15 μg/ml β-galactosidase.
<u>2nd day:</u> 90 small test tubes. 1 empty test tube rack for 90 tubes. 150 ml of 0.03 M phosphate buffer, pH 7.4. 4 ml of 2-nitrophenyl-β-D-galactopyranoside solution (ONPG-solution 5 mg/ml). 3 ml of 0.01 M sodium pyruvate solution. 3 ml of 0.002 M NADH, disodium salt, freshly prepared. For several groups together: 1 photometer. 1 gradient dripping device with a hypodermic needle of 0.6 mm inner diameter (see p. 41).

<u>Procedure</u>

<u>Preparation of gradient (1st day).</u> Close the connecting stop cock of the gradient mixing vessel. Pipette 3.7 ml of a 5% sucrose solution into the chamber having no outlet and 3.7 ml 20% sucrose solution into the chamber with the outlet. Put the stirrers into both chambers, so that the menisci are level, but only stir in the chamber having the outlet. Then open the connecting stop cock between both chambers. Interchangeably, fill the sucrose solution into two centrifuge tubes drop by drop. This way two groups can form their gradients simultaneously with one mixer. A small slice of cork, which is placed at the

bottom of each centrifuge tube before the gradient is formed,
prevents turbulences by the falling drops. After the centrifuge tubes have been filled, carefully remove the slices of
cork with a pair of forceps. Layer 0.15 ml of the sample solution on top of the gradient, using a PASTEUR pipette. Balance
the centrifuge tubes of the groups with paraffin and centrifuge
overnight (approx. 15 hrs) in an SW56Ti rotor at 30,000 rpm
(90,000 × g) at 4°C.

Dripping of the gradient (2nd day). Pierce the bottom of the
centrifuge tube by means of the dripping device and collect
fractions of 8 drops. Collect a total of about 35 fractions.

Evaluation of the Gradient

Cytochrome c: Add 1.5 ml of 0.03 M phosphate buffer, pH 7.4,
to each of the fractions 29-35, mix, and measure $O.D._{420}$ against
phosphate buffer as a reference.

Lactate dehydrogenase (LDH): Add 0.3 ml of 0.03 M phosphate buffer,
pH 7.4, to each of the fractions 18-28. Prepare 12 test tubes with

    0.1 ml sodium pyruvate solution

    0.1 ml NADH-solution.

Then pipette, successively, 2.7 ml of 0.3 M phosphate buffer at
37° and, at time t = 0, 0.1 ml of the diluted LDH fractions.
Incubate at 37°. Stop the reaction at time t = 45 sec, by
placing the reaction mixture into a boiling water bath for
7 min. Finally cool the samples to room temperature. Treat all
11 LDH fractions successively in this manner. As a blank use
the same reaction mixture as for the LDH test but replace the
LDH solution by the same amount of phosphate buffer. Read the
samples photometrically. Adjust the blank at $O.D._{340}$ to 0.4
and measure the O.D. differences to this value for all other
samples.

β-Galactosidase: Add 1.5 ml of 0.03 M phosphate buffer, pH 7.4,
to each of the fractions 1-17 and mix. Take 0.5 ml from each

    Add   4 ml phosphate buffer at 37°C
    add 0.2 ml ONPG solution

Keep at 37°C in a water bath for 10 min and then immediately
read the $O.D._{420}$ against a blank which contains 0.5 ml phosphate
buffer instead of 0.5 ml β-galactosidase solution.

Record all O.D. readings graphically:

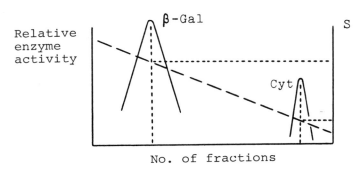

No. of fractions

For the <u>graphical evaluation</u>, use the right ordinate and the di-
agonal standard curve, whose slope is determined by the given
s-values and the position of the maxima of activity of the reference molecules.

The <u>mathematical evaluation</u> requires the use of formula (2)
given above:

Reference molecule	Reference s-value	$\dfrac{D_{LDH}}{D_{reference}}$	s_{LDH}
Cytochrome c	1.7 S		
β-Galactosidase I	15.9 S		
β-Galactosidase II (a second maximum)	23 S		

The 23 S maximum of β-galactosidase does not appear in all preparations of β-galactosidase.

Literature

MAHLER, H.R., CORDES, E.H.: Biological Chemistry. Theory of Ultra-
    centrifugation, see pp. 80 ff, 218 ff. New York – London: Harper
    & Row 1971.

SOBER, H.A. (ed.): Handbook of Biochemistry. Selected Data for
    Molecular Biology. See Sections C10, H4 (S-values). Cleveland:
    CRC Press 1970.

SUND, H., WEBER, K.: Untersuchungen über milchzuckerspaltende
    Enzyme, XIII. Größe und Gestalt der β-Galactosidase aus *E. coli*.
    Biochem. Zeitschr. <u>337</u>, 24-34 (1963).

Time requirement: 1st day 1 hr, 2nd day 4 hrs.

Gradient mixer

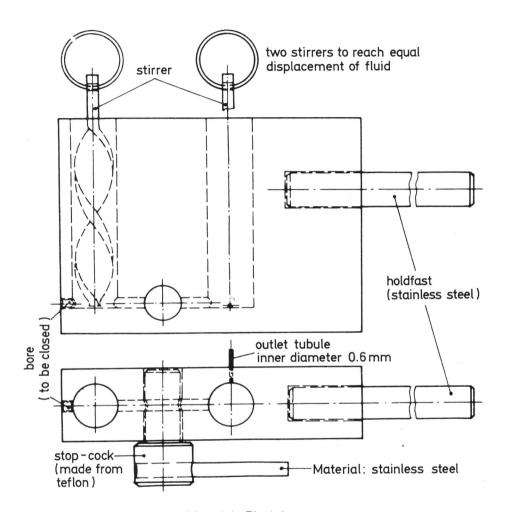

stirrer

two stirrers to reach equal
displacement of fluid

holdfast
(stainless steel)

bore
(to be closed)

outlet tubule
inner diameter 0.6 mm

stop-cock
(made from
teflon)

Material: stainless steel

Material: Plexiglass

48

Data sheet

Fraction No.	O.D.$_{420}$ (Cytochrome c)	ΔO.D.$_{340}$ (LDH)	O.D.$_{420}$ (β-Gal.)
1 2 3 4 5			
6 7 8 9 10			
11 12 13 14 15			
16 17 18 19 20		– –	– – –
21 22 23 24 25			
26 27 28 29 30	– – –	– –	
31 32 33 34 35			

Problems

1. If a phage infects a bacterium and no progeny are formed or set free, this might have the following reasons:

   a) lysogenization

   b) restriction

   c) host immunity

   d) non-permissive growth conditions

Define expressions (a) to (d) and give an example of each.

2. Characterize the DNA of the following phages by indicating (+) and (-) in the table.

Phage	The DNA molecule is				The base sequences has		
	linear	circular	single stranded	double stranded	terminal redundancy	cyclic permutation	cohesive ends
T2, T4 T7							
λ extracell. λ intracell.							
ΦX174, fd extracell. intracell.							

extracellular = DNA isolated from phage particles
intracellular = DNA isolated from infected bacteria

3. The sedimentation constant and the density of macromolecules can be determined by ultracentrifugation. In an analytical ultracentrifuge the distribution of the molecules in the centrifugation tube can be observed optically before, during and after centrifugation. Which two types of centrifugation are represented in the graphs below and how are they characterized.

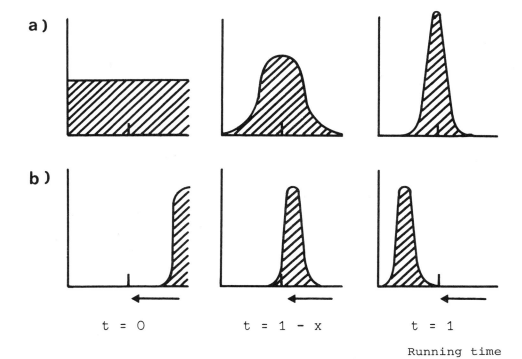

a)

b)

t = 0   t = 1 - x   t = 1

Running time

B. Nucleic Acids and Transcription

Although nucleic acids were isolated and described in the last
century (MIESCHER), their key position among cell components
has been recognized in only the past 30 years:

- AVERY and co-workers (1944) proved, by transformation experiments, that deoxyribonucleic acid (DNA) is the carrier
  of genetic information. As life without enzymes and other proteins is impossible, Avery's discovery implied that there must
  be a connection between DNA and the synthesis of protein. The
  question of how information for protein synthesis is coded in
  DNA and is then passed on could only be answered after the
  chemical structure of DNA was known.

- In 1953 WATSON and CRICK published a model for the macromo-
  lecular structure of DNA. According to this DNA is an unbranched, linear, double-stranded macromolecule, consisting
  of two helically-arranged polynucleotide chains with complementary base sequences (Expt. 6). This molecule is replicated
  semi-conservatively. So far, three different DNA polymerases
  have been described, of which DNA polymerase III is the most
  important. However, some other enzymes and proteins are also
  involved in DNA replication.

- Towards the end of the fifties, a hitherto unknown type of
  ribonucleic acid (RNA) was found in T2-infected cells. It has
  an average base composition, similar to that of T2 DNA and is
  an unstable cell component. Moreover, an enzyme, "DNA-dependent
  RNA polymerase", was discovered which synthesizes RNA with a
  base sequences complementary to that of a given DNA template.
  It was presumed that this RNA is the "messenger" for genetic
  information on its way to protein synthesis and therefore it
  was termed messenger RNA (mRNA).

$$\text{DNA} \xrightarrow{\substack{\text{RNA Polymerase} \\ \text{+ Triphosphates}}} \text{mRNA} \xrightarrow{\substack{\text{Ribosomes + tRNAs} \\ \text{+ several co-factors}}} \text{Protein}$$

The *in vitro* synthesis of specific proteins in the presence of
natural or synthetic mRNA molecules finally proved the cor-
rectness of this hypothesis.

Genetic experiments with phage T4 had shown that genetic information is coded by means of a triplet code in the nucleic acid: a
sequence of three nucleotides specifies one amino acid and thus
linear sequences of polynucleotides determine linear sequences
of amino acids in polypeptides. Using synthetic mRNA with known
base sequences it was soon possible to show which base triplet
codes which amino acid.

Expt. 6 deals with the isolation of phage DNA. The average GC-content and the double strandedness of this DNA can be deduced from Expt. 7, which demonstrates the thermal denaturation and renaturation of DNA. The extraction of radioactively labeled phage specific mRNA and its hybridization with single-stranded DNA is shown in Expts. 8 and 9. The last experiment in this section proves that RNA is also a carrier of genetic information and that the RNA of individual genes is not synthesized continuously.

Literature

CANTONI, G.L., DAVIES, D.R. (eds.): Procedures in Nucleic Acid Research, 667 p. London: Harper and Row Publ. 1966 and Vol. 2, 924 p. (1971).

CHARGAFF, E., DAVIDSON, J.N. (eds.): The Nucleic Acids. 3 Volumes. New York: Academic Press 1955-1960.

DAVIDSON, J.N.: The Biochemistry of the Nucleic Acids, 396 p. London: Chapman and Hall 1972.

DAVIDSON, J.N., COHN, W.E. (eds.): Progress in Nucleic Acid Research and Molecular Biology. 14 Volumes. London: Academic Press 1963-1974.

GROSSMAN, L., MOLDAVE, K. (eds.): Methods in Enzymology, Vol. 21. Nucleic Acids, Part D. London: Academic Press 1971.

KORNBERG, A.: DNA Synthesis, 399 p. San Francisco: Freeman and Co. 1974.

MOLDAVE, K., GROSSMAN, L. (eds.): Methods in Enzymology, Vol. 20. Nucleic Acids and Protein Synthesis, Part C. London: Academic Press 1971.

6. Isolation of DNA from T4 or T5 Phages

Deoxyribonucleic acid (DNA) is a macromolecule of very high molecular weight (MW). For example, the MW of phage DNA ranges between
10^6 and 10^8 daltons. The molecular weight of the DNA of *E. coli*
is about 3×10^9 daltons. The basic unit of a DNA-molecule is
the deoxyribonucleotide, which is composed of three subunits:
one molecule each of purine bases (adenine and guanine) or of
pyrimidine bases (thymine and cytosine) attached by an N-glycos-
idic bond to one molecule of deoxyribose to which is bound, in
its 5' position, a phosphoric acid residue. Each nucleotide with
its phosphoric acid residue is linked to the 3' position of the
deoxysugar of the preceding nucleotide, thus forming an unbranched
linear polynucleotide (primary structure).

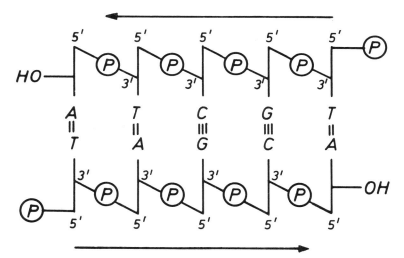

A double-stranded DNA molecule is composed of two of these polynucleotide chains with complementary base sequences: a purine
base in one of the chains faces a pyrimidine base in the other
chain and vice versa. The two polynucleotide chains are held together by hydrogen bonds between the purine and pyrimidine
bases. Generally stable hydrogen bonds only form between the
"complementary" bases adenine and thymine (A-T) and guanine and
cytosine (G-C). Therefore, the molar ratios of adenine to thymine
and of guanine to cytosine are equal to one in double-stranded
DNA since the base sequence of one DNA strand is determined by
the base sequence of the other. As can be seen from the schematic
presentation above, the 3',5' internucleotide phosphodiester
bridges of both polynucleotide chains have antiparallel polarity:
in one strand they run in the 3'⟶5' direction and in the other
in the 5'⟶3' direction. When considering this schematic model
it should be kept in mind that double-stranded DNA actually has
a helical three-dimensional structure. The sugar-phosphate chains
coil around a common axis. The purine and pyrimidine bases are
stacked on the inside of the double helix with their planes

parallel to each other and at a right angle to the helix axis (secondary structure). The double helix is stabilized by hydrogen bonds between the complementary base pairs and by hydrophobic interaction between the stacked bases.

Native double-stranded DNA is sensitive to shearing and other forces. Therefore care has to be taken not to reduce its molecular weight by pipetting a DNA solution through capillaries or by exposing it to shearing forces as by vigorous shaking.

Different exo- and endonucleases hydrolyze DNA. The hydrogen bonds between the bases are opened by heating of the DNA (see Expt. 7), by decreasing the ionic strength or by treatment with alkali (pH 10) or acid (pH 3). At low pH, the N-glycosyl linkages between the deoxyribose and the purine residues are split, resulting in an apurinic acid.

The extraction of DNA from phages is quite easy, because in phages the ratio of protein to DNA is about 1:1, whereas in bacteria and other cells protein dominates. In addition, in phages there are no cellular nucleases nor polysaccharidic cell wall components to complicate the DNA extraction. The DNA content of phage T4 is 2.2×10^{-16}g (MW 1.3×10^8), and that of phage T5 is 1.4×10^{-16}g (MW 8.5×10^7). The T-even phages (T2, T4, T6) contain 5-hydroxymethyl cytosine (HMC) instead of cytosine. In addition, one or two glucose molecules may be linked to the hydroxymethyl groups of the HMC.

Plan. DNA will be isolated from T4 or T5 phages by phenol extraction and will be dialyzed against a 1:100 diluted standard saline citrate-solution. The amount of DNA isolated will be determined photometrically.

Material. 10 ml of a T4 or T5 phage suspension (approx. 1×10^{12} phages/ml). About 30 ml of phenol, freshly distilled, saturated with 0.1 M tris-HCl buffer, pH 7.2. 3 50-ml centrifuge tubes. 1 rubber stopper. Parafilm. Disposable plastic gloves. 4 PASTEUR pipettes. Dialysis tubing. Cotton. 2 2-l Erlenmeyer flasks. 5 ml 20-times concentrated SSC (20 x SSC) for preparation of the dialysis buffer. Magnetic stirrer and stirring motor. For several groups together: UV-spectrophotometer. SORVALL centrifuge RC2-B with rotor SS34.

Procedure

Since phenol causes burns, when working with phenol, wear safety goggles, use plastic gloves and pipette only with a pipette bulb or cotton plugged pipettes.

Pipette 10 ml of the phage suspension and 10 ml of cold phenol (4°C) into a centrifuge tube. Close the centrifuge tube with a rubber stopper, which has been covered with parafilm, and carefully (to avoid shearing forces) tilt back and forth for 10 min. Remove the stopper and dry the edge of the tube with paper. Then, centrifuge at 12,000 × g (approx. 10,000 rpm) for 3 min. The aqueous upper phase and the phenolic lower phase are separated by this treatment. The denatured phage protein is seen as a very thin, white interphase. Use a PASTEUR pipette with the tip cut off in order to obtain a wide pipette opening and thus avoid excessive shearing. Carefully pipette the aqueous phase, which contains the DNA, into a second centrifuge tube. Avoid, as far as possible, pipetting the protein interphase. Again add 10 ml of cold phenol and treat as before. Repeat this procedure a third time. After the last centrifugation, transfer the aqueous phase, free from protein and phenol, into dialysis tubing* and dialyze overnight against 2 l 0.01 × SSC each. Change dialysis buffer 3 times.

The dialyzed DNA is kept in a sterile test tube. A few drops of chloroform are added in order to keep the DNA sterile. Pipette 0.1 ml of this solution to 0.9 ml 0.01 × SSC and mix. Determine the amount of DNA at $O.D._{260}^{1\,cm}$. An $O.D._{260}^{1\,cm}$ of 1.0 corresponds to a DNA concentration of about 50 µg/ml.

Appendix: The preceding experiment can be modified. Instead of dialysis against 0.01 × SSC, the DNA can also be precipitated with ethanol: make the aqueous phase 1% by the addition of a 20% solution of potassium acetate. Carefully overlayer the DNA solution with two volumes of ice cold 96% ethanol. The DNA precipitates at the interphase between the alcoholic and the aqueous layers. The DNA can then be collected on a glass rod which is slowly rotated in the aqueous phase. The DNA is dried by transferring the glass rod successively into three separate tubes containing 10 ml of 96% ethanol (10 min each). Finally, the ethanol is removed by acetone (procedure as described above for ethanol) and the DNA is air dried).

Literature

THOMAS, C.A., Jr., ABELSON, J.: The Isolation and Characterization of DNA from Bacteriophage. In: Procedures in Nucleic Acid Research (eds. J.L. CANTONI, D.R. DAVIES), p. 553-561. New York: Harper & Row 1967.

WATSON, J.D.: The Double Helix, 183 p. London: Weidenfeld and Nicolson 1968.

Time requirement: 1st day 3 hrs, 2nd day 0.5 hrs.

---

* The dialysis tubing should be previously boiled in 5% Na_2CO_3 2 × 15 min, washed in distilled water and finally boiled for 15 min in 0.05 M Na_3EDTA and washed in distilled water.

7. Thermal De- and Renaturation of DNA from Different Organisms

The DNA of all organisms and many viruses is, in its native state, a double-stranded molecule (see Expt. 6). The hydrogen bonds between the complementary bases A-T and G-C as well as the stacking forces between the neighboring bases are responsible for the double helical structure of DNA and its stability. If a solution of native DNA is heated to $100^{\circ}C$ in neutral buffer or if it is made highly alkaline, the "ordered" structure collapses, i.e., the DNA becomes single-stranded and the single strands form random coils. The lower the guanine and cytosine content of the DNA, the easier this denaturation occurs. Denaturation is accompanied by an increase in the optical density measured at 260 nm. The physical basis for this "hyperchromic effect" is the fact that polynucleotides, when in the ordered state of a double-stranded DNA molecule, absorb less light of short wave length than would be expected of an equal quantitiy of mononucleotides of the same molar ratios. Double-stranded DNA therefore shows "hypochromicity". A possible explanation for this effect at the molecular level might be that excitations of the π-electrons of the bases (the electrons, which absorb the light) are strongly reduced in the long stacks of nucleotides in a native DNA molecule.

When the optical density of a heated DNA solution is plotted against the temperature, a "melting curve" is obtained. The melting point T_m (measured in $^{\circ}C$) of this curve is directly proportional to the GC-content and furthermore depends on the ionic strength of the DNA solution. If DNA which has been heat-denaturated is slowly cooled, the complementary base sequences have time to rearrange, i.e., the DNA renatures. When chilled quickly, renaturation is rarely observed. De- and renaturation are important methods to characterize nucleic acids (see Expt. 9).

Plan. Two DNA solutions of different origin are heated from 40° to $100^{\circ}C$. During this procedure, the O.D. is measured photometrically at 260 nm. The $O.D._{260}$ values will be plotted as a function of the temperature. The melting point, the average GC-content and the hyperchromicity will be determined from this curve. After slowly cooling the denatured DNA, the percentage of DNA which has renatured under the given conditions will be estimated.

Material. 10 ml of standard saline-citrate (SSC) solution. 3 ml of DNA (approx. 20 µg/ml) dissolved in SSC, e.g., DNA from *E. coli* BA and salmon sperm. For several groups together: 1 UV spectrophotometer with temperature controlled cuvette holder. 1 thermostat with pump. 1 fine thermo-element attached to a potentiometer standardized for the range of 40-100°C, to serve as a temperature gauge. 4 quartz cuvettes (1 cm light path). 1 standard curve: melting point plotted against GC-content (see p. 60).

Procedure

1. <u>Preparation:</u> Put in cuvette holder:

   Cuvette 1: 3 ml SSC with thermoelement

   Cuvette 2: 3 ml SSC

   Cuvette 3: 3 ml *E. coli* - DNA in SSC

   Cuvette 4: 3 ml salmon sperm-DNA in SSC

Put lids on all cuvettes in order to prevent evaporation; a lid with a fitting for the thermoelement must be constructed. Set the thermostat at 40°C, using the contact thermometer and turn on the pump.

2. <u>Denaturation:</u> Read the potentiometer at time t = O and determine the optical density (O.D.) of cuvettes 3 and 4 at 260 nm; cuvette 2 serves as a blank. Then set the contact thermostat at 100°C. While the temperature is slowly rising to 100°C (approx. 1 hr), record the $O.D._{260}$ values and the potentiometer readings at intervals of 2 min. Continuously control the zero point setting with cuvette 2. Push the cuvette holder carefully in order not to damage the thermoelement.

3. <u>Renaturation:</u> If the water bath in the ultrathermostat begins to boil, immediately turn off the heater, but keep pump going. Leave all the cuvettes in the cuvette holder. While the water bath slowly cools down, part of the denatured DNA will renature. Read the $O.D._{260}$ values and the potentiometer values at intervals of 30 min for 3-4 hrs.

Evaluation

1. Convert the potentiometer readings to temperature with the aid of a thermoelement standard curve (see data sheet). Plot the $O.D._{260}$ values graphically against the temperature. Connect all points obtained before, during and after the $O.D._{260}$ shift as indicated in the figure. The approximate temperature interval (Δ_t) at which the DNA under investigation melts is obtained from the intersecting points of these horizontals with the melting curve.

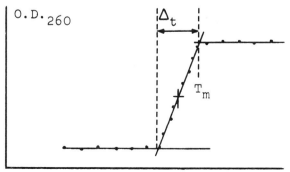

The melting point, T_m, lies at approx. $1/2 \, \Delta_t$. The GC-content can be read from the GC standard curve, which corresponds to different T_m values.

    E. coli DNA: T_m = ... °C   GC-content ...... %

    Salmon sperm DNA: T_m = ... °C   GC-content ...... %

2. Calculate the <u>hyperchromicity</u>: divide the average O.D.$_{260}$ value after denaturation by the corresponding value before denaturation. Subtract 1 from this ratio.

    E. coli DNA: ————— - 1 = _____

    Salmon sperm DNA: ————— - 1 = _____

3. Estimate the <u>extent of renaturation</u>: At 260 nm

$$\frac{(\text{O.D.}_{before} - \text{O.D.}_{after\ renaturation}) \times 100}{(\text{O.D.}_{after} - \text{O.D.}_{before\ denaturation})} = \text{...... \% (\textit{E. coli} DNA)}$$

$$= \text{...... \%}$$
(Salmon sperm DNA)

Literature

HILL, L.R.: The Determination DNA Base Compositions and its Application to Bacterial Taxonomy. In: Identification Methods for Microbiologists, Part B, pp. 177-186. London: Academic Press 1968.

LAZURKIN, YU.S., FRANK-KAMWNETSKII, M.D., TRIFONOV, E.N.: Melting of DNA: Its Study and Application as a Research Method. Biopolymers 9, 1253-1306 (1970).

MARMUR, J., ROWND, R., SCHILDKRAUT, C.L.: Denaturation and Re-naturation of Deoxyribonucleic Acid. In: Progress in Nucleic Acid Research, Vol. 1, pp. 231-300. New York: Academic Press 1963.

MARMUR, J., SCHILDKRAUT, C.L., DOTY, P.: Biological and Physi-cal Chemical Aspects of Reversible Denaturation of DNA. In: The Molecular Basis of Neoplasia, pp. 9-43. Austin: University of Texas Press 1962.

<u>Time requirement:</u> 1st day 5 hrs.

Data sheet

Time	Potentio-meter reading	Temperature °C	Extinction at 260 nm	
			E. *coli* DNA	Salmon sperm DNA
Denaturation (min):				
0				
2				
4				
6				
8				
10				
12				
14				
16				
18				
20				
22				
24				
26				
28				
30				
32				
34				
36				
38				
40				
42				
44				
46				
48				
50				
52				
54				
56				
58				
60				
Renaturation (hrs):				
0				
0.5				
1.0				
1.5				
2.0				
2.5				
3.0				
3.5				
4.0				

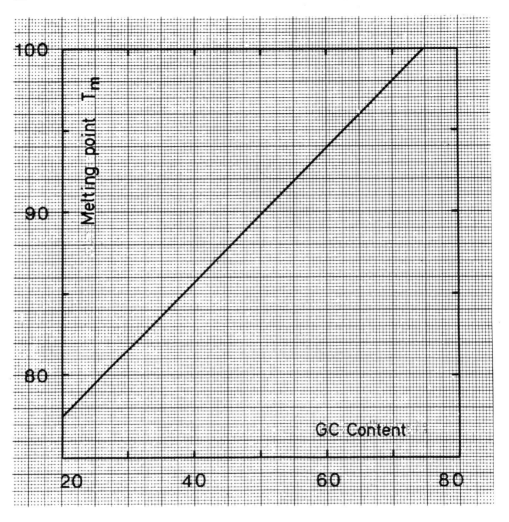

The melting point (°C) of DNA as function of GC-content (%),
calculated as follows: $GC = \dfrac{T_m - 69.3}{0.41}$ (see L.R. HILL, 1968).
This relation is only valid for DNA dissolved in SSC. For other
ionic strengths see formula on p. 192

8. Isolation of ³H-Uridine-Labeled mRNA from Bacteria Infected with T4

DNA-dependent RNA polymerase transcribes the genetic information
from DNA into messenger RNA (mRNA). When double-stranded DNA
serves as a template "in vivo" (see figure), the transcribed
RNA of a certain group of genes, e.g. A, is complementary to
only one of the two DNA strands. The messenger is transcribed
asymmetrically and we can differentiate "sense" and "antisense"
information in DNA.

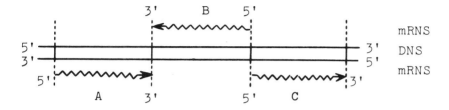

However, the mRNA transcript of a second group of genes, e.g. B,
can be complementary to the other strand, i.e., both strands of
DNA can carry "sense" as well as "antisense" information.

RNA polymerase reads the information on the DNA molecule in
the 3'⟶5' direction and synthesizes RNA molecules in 5'⟶3'
direction using 5' nucleoside triphosphates as substrate
(transcription).

The molecules of mRNA transfer genetic information from the DNA
to the ribosomes (messenger RNA). The ribosomes then synthesize
polypeptide chains with amino acid sequences corresponding to the
base sequences of the mRNA (translation). The translation of mRNA
also runs in the 5'⟶3' direction. Here the 5' end of mRNA corresponds to the amino terminal end of the polypeptide.

DNA and mRNA differ as follows:

DNA	mRNA
Deoxyribose	Ribose
Thymine	Uracil
Double strand	Single strand
MW 10^6–10^9	MW 10^5–10^6

In cells DNA is stable, whereas mRNA is continuously degraded
and resynthesized. The average "half-life" of mRNA molecules in
microorganisms is about 90 sec.

Plan. *E. coli* will be infected with T4 phages in the presence of ^{3}H-uridine. 10 min after infection further phage development will be stopped (1st day) and T4 specific mRNA will be isolated from the bacteria (2nd day).

Material. 1st day: 10 ml of a stat culture of *E. coli* BA in M9 (37°C, aerated). 100 ml of minimal medium M9, supplemented with 2 mM MgCl$_2$ and 0.01 mM FeCl$_3$. 2.5 ml of a suspension of T4 phages (wild type) having a titer of 1×10^{11}/ml. 1.5 ml of a tryptophan solution (100 mg/ml). 40 ml of frozen and 15 ml of ice-cold sodium azide buffer (10 mM tris HCl, pH 7.3, 5 mM MgCl$_2$, 10 mM sodium azide). 1 250-ml centrifuge tube. 1 12-ml centrifuge tube. Parafilm. For several groups together: 1 SORVALL centrifuge RC2-B with rotor GSA, 100 µCi of ^{3}H-uridine (specific activity 1,000-3,000 mCi/mmole). Methanol bath -60°C or a methanol-dry ice mixture.
2nd day: 50 µl of lysozyme, 30 mg/ml. 30 µl of DNase (RNase-free), 1 mg/ml. Both enzyme solutions should be freshly prepared and kept in ice. 0.5 ml of 20 mM acetic acid. 0.3 ml of a 10% sodium dodecyl sulfate solution. 5 ml of phenol, freshly distilled over zinc granules and equilibrated with 10 mM potassium acetate buffer, pH 5.2. 30 ml of SEPHADEX G25, coarse, equilibrated with 10 mM potassium acetate buffer, pH 5.2. 30 ml of DOWEX 50W-X8, 20-50 mesh, counter ion K$^+$ (obtained by washing in 0.5 N KOH, followed by several washes with distilled water. The resin is equilibrated in 10 mM potassium acetate buffer, pH 5.2). Disposable plastic gloves. 2 PASTEUR pipettes. 2 20-µl micropipettes and 1 40-µl micropipette (Corning). Parafilm. 1 rack with 60 small test tubes. 1 50-ml centrifuge tube with rubber stopper to fit. Cotton. 1 stand with 1 clamp and 1 chromatography column (1.5 x 40 cm). 50 ml of 96% ethanol, ice-cold. About 300 ml of dioxane scintillation fluid. 30 PACKARD scintillation vials. For several groups together: 1 water bath at 20°C. 1 UV-spectrophotometer. 1 PACKARD scintillation counter. 1 SORVALL centrifuge RC2-B with rotor SS34. 1 37°C water bath.

Procedure

When working with radioactive materials, safety measures, which are required by law, must be observed at all times. Use only pipette bulbs or similar safety devices to pipette radioactive solutions. If the entire 100 µCi of ^{3}H-uridine are used, the experiment should be done by only one group under the supervision of the laboratory instructor. Alternatively, each group may use only 10 µCi of ^{3}H-uridine per experiment.

1st day: Inoculate 100 ml of M9 minimal medium with 10 ml of the stat culture of *E. coli* BA and aerate at 37°C. The actual experiment begins about 3 hrs later, when the cell density reaches 5×10^8/ml, corresponding to an O.D.$^{1cm}_{660}$ of 0.5.

t = 0 min   Add 100 µCi ^{3}H uridine

t = 4.5 min Add 1 ml tryptophan solution (needed as a co-
             factor for the adsorption of T4 phages).

t = 5 min   Add 2 ml T4 phages (multiplicity of infection = 4).

t = 10 min   Quickly cool culture by transferring it to a 250-ml centrifuge tube, which contains 40 ml frozen, crushed Na-azide buffer. (Na-azide prevents any further development of phage by uncoupling phosphorylation and inhibiting iron-containing enzymes).

Centrifuge at 6,000 rpm (approx. 6,000 × g) in the GSA rotor for 3 min. Discard supernatant – transfer radioactive supernatants into radioactive waste container! Resuspend the pellet in ice-cold Na-azide buffer and centrifuge again. Discard supernatant. Resuspend the pellet in 2 ml ice-cold Na-azide buffer (PASTEUR pipette), freeze in a small centrifuge tube in the methanol bath at -60°C and seal with parafilm. Keep in the deep freezer (-20°C).

<u>2nd day</u>: Cell Disintegration and Phenol Treatment
Thaw suspension of frozen bacteria in a 20°C water bath. Keep at this temperature and treat as follows:

O min   Add of 40 µl of the <u>lysozyme</u> solution, to lyse the cell walls (final concentration: 600 µg/ml).

2 min   Add 20 µl of <u>deoxyribonuclease</u>, to degrade the DNA (final concentration: 10 µg/ml).

4 min   Add 0.2 ml of 20 mM <u>acetic acid</u>, to lower the pH of the solution to about pH 5. Add 0.1 ml of 10% <u>sodium dodecyl sulfate,</u> to further disintegrate the cells (final concentration: 5 mg/ml). Finally, incubate at 37°C for 3 min. The turbid cell suspension should clear.

8 min   Add 2 ml of ice-cold phenol.

Seal centrifuge tubes with parafilm and rubber plug and shake well for 5 min. Remove rubber plug and dry rim of tube. Centrifuge at 12,000 × g for 5 min (SS34 rotor, 10,000 rpm). RNA is contained in the aqueous upper phase. The denatured protein appears in the interphase.

Column Chromatography

Place the aqueous (protein-free) phase on a chromatography column using a PASTEUR pipette. The lower half of this column contains 30 ml SEPHADEX G25, the upper half 30 ml of the DOWEX ion exchange resin carefully layered on top of the G25 resin. The DOWEX absorbs the basic proteins and, by gel-sieve chromatography, the SEPHADEX then separates the RNA from the phenol and other low molecular weight substances. Equilibrate the column with 10 mM potassium acetate buffer, pH 5.2. Pass the sample through the column with the same buffer at a rate of about 3 ml/min. Collect 30 fractions of 2.5 ml each.

Measurement of Absorption and Radioactivity

Dilute fractions 1:10 in potassium acetate buffer, pH 5.2 (0.2 ml
sample + 1.8 buffer) and take O.D. readings from each diluted
fraction

a) at 260 nm (λ_{max} RNA) and at 270 nm (λ_{max} phenol) in a UV
   spectrophotometer,

b) pipette 20 µl (rinse pipette each time) to 5 ml dioxane
   scintillation fluid and measure the radioactivity in a
   liquid scintillation counter.

Both the $O.D._{260}$ measurements and the radioactivity measurements
show the fractions containing the RNA. Pool the 5-7 undiluted
fractions, which contain the main portion of RNA, in a 50-ml
centrifuge tube, adjust to 1% with 20% potassium acetate and
mix with 30 ml of ice-cold ethanol. RNA precipitates as the
potassium salt. In order to precipitate all the RNA from the
mixture, leave the tube in ice for 30 min and then centrifuge
at 35,000 × g (SS34 rotor, 17,000 rpm) for 15 min. Save RNA
pellet and discard the supernatant.

If the RNA is to be used immediately (e.g., for Expt. 9), dissolve in 0.01 × SSC. If not, it can be preserved under alcohol
at -20°C for several months.

Evaluation

Plot the optical density graphically against fraction number
(see data sheet).

1. Estimate the amount of RNA isolated:

$$\sum_{i=1}^{n} (O.D._{260})_i \times 2.5 \text{ ml} \times 10 \times 42 \times 10^{-3} = \underline{\qquad} \text{ mg RNA.}$$

Use 5-7 peak fractions (i). The multiplication factors,
$10 \times 42 \times 10^{-3}$ result from the 1:10 dilution for the O.D.
reading and the fact, that 1 $O.D._{260}$ unit corresponds to
42×10^{-3} mg RNA/ml.

2. Calculate the specific activity of the isolated RNA using
the data of the peak fraction, although all other fractions
have the same specific activity.

The radioactivity was measured from those RNA fractions which
were diluted 1:10 taking 20 µl-samples each time. The following
results were obtained:

$$(\text{.... cpm/20 µl}) \times 50 \times 10 = \underline{\qquad} \text{cpm/ml in the undiluted RNA fraction}$$

To convert cpm to dpm, multiply by 2, because under the given conditions (^{3}H) the count yield is only about 50% of the disintegrations/per min (dpm) which actually occur.

......(cpm/ml) × 2 = _____ dpm/ml.

1 mCi corresponds to 2.2×10^9 dpm. The fractions therefore contain

$$\frac{dpm/ml}{2.2 \times 10^9} = \text{_____ mCi/ml.}$$

Now this radioactivity should be expressed as per mg RNA in the fraction measured, in order to obtain the "specific activity".

O.D.$_{260}$ × 10 × 42 × 10^{-3} = _____ mg RNA/ml.

The specific activity is obtained by equating the radioactivity with the content of RNA:

$$\frac{mCi/ml}{mg\ RNA/ml} = \text{_____ mCi/mg RNA .}$$

Literature

BAUTZ, E.K.F., HALL, B.D.: The Isolation of T4-Specific RNA on a DNA-Cellulose Column. Proc. Nat. Acad. Sci. U.S. __48__, 400-408 (1962).

MAGASANIK, B.: Isolation and Composition of the Pentose Nucleic Acids and of the Corresponding Nucleoproteins. In: The Nucleic Acids (eds. E. CHARGAFF, J.N. DAVIDSON), Vol. I, pp. 307-368. New York: Academic Press 1955.

Special Radioisotope Literature

CHASE, G.D., RABINOWITZ, J.L.: Principles of Radioisotope Methodology. Minneapolis: Burgess Publ. Comp. 1968.

WOLF, G.: Isotopes in Biology. London: Academic Press 1964.

Time requirement: 1st day 4 hrs, 2nd day 5 hrs.

Data sheet

Fraction No.	O.D.$_{260}$	O.D.$_{270}$	$\dfrac{\text{O.D.}_{260}}{\text{O.D.}_{270}}$	Radioactivity cpm/20µl
1				
2				
3				
4				
5				
6				
7				
8				
9				
10				
11				
12				
13				
14				
15				
16				
17				
18				
19				
20				
21				
22				
23				
24				
25				
26				
27				
28				
29				
30				

9. Hybridization of Phage-DNA and Phage-mRNA on Nitrocellulose Filters

If double-stranded DNA is heated to 90°C in a solution of low salt concentration, the hydrogen bonds between the bases open and the DNA decomposes into single strands. If such a solution of denatured DNA is slowly cooled, part of the single-stranded DNA renatures, due to complementary base sequences in the DNA double strands (compare Expt. 7). If denatured DNA of different origins is mixed, e.g., from phage T3 and phage T7, and if renaturation conditions are maintained, hybrid molecules consisting of one T3 and one T7 strand, as well as T3 and T7 double helices are formed. Hybridization proves that DNA of both hybridization partners has identical or very similar base sequences. Thus the formation of hybrid molecules is a measure of the "relationship" of nucleic acid molecules. RNA can also hybridize with complementary DNA or RNA. Depending on the hybridization partners one can refer to the processes of DNA-DNA, DNA-RNA or RNA-RNA hybridization.

The most widespread of the hybridization techniques is DNA-RNA hybridization. First, single-stranded DNA is immobilized on nitrocellulose filters, so that it cannot renature. If radioactively labeled mRNA then is added to the "DNA-filter", radioactivity is retained on the filter only when RNA has formed hybrid molecules with the DNA. In contrast to single-stranded DNA, RNA does not stick to nitrocellulose. Therefore, the radioactivity detected on the filters is a direct measure of the amount of hybridized RNA. If DNA is to be hybridized to the DNA on the filter, the remaining sites for DNA attachment on the nitrocellulose filter must first be saturated with serum albumin before the addition of the radioactively labeled DNA. Then radioactivity bound to the filter is a measure of the hybridized DNA.

The amount of hybridization of two nucleic acids is not only dependent on their base sequences, but also on the molecular weight, the ionic strength and the temperature at which the hybridization takes place. As RNA does not bind to nitrocellulose, RNA-RNA hybridization is carried out in solution. Subsequently, non-hybridized RNA is broken down with pancreatic ribonuclease; contrary to RNA single strands, double-stranded RNA is not hydrolyzed and therefore remains acid-precipitable. Thus, the percentage of RNA hybridized can be directly determined.

Plan. RNA from bacteria infected by T4 phages (Expt. 8) will be shown to hybridize only with DNA of T4 phages and not with DNA of T5 phages.

Material. 15 nitrocellulose filters (SCHLEICHER & SCHÜLL BA 85, 25 mm diameter, pore size 0.45 μm). 15 ml of a solution of T4 DNA or T5 DNA (Expt. 6), adjusted to 100 μg/ml in 0.01 × SSC. About 0.15 ml of ^{3}H uridine-labeled mRNA (approx. 16 O.D. units/ml), isolated from *E. coli* bacteria infected with phage T4 (Expt. 8), specific activity >1.5×10^6 cpm/mg. 1,000 ml of 2 × SSC. 20 ml of 20 × SSC. 30 PACKARD scintillation vials. Tweezers. Paper hand towels. 1 20-µl pipette. 1 50-ml graduated cylinder. 100 ml of toluene scintillation fluid. 5 100-ml beakers. For several groups together: Water pressure pump. Filter box for 10 filters (see p. 71). 1 vacuum pump. 1 vacuum oven. 1 scintillation counter (PACKARD). 1 65°C water bath. 1 250-ml Erlenmeyer flask. 5 100-ml beakers.

Procedure

1. <u>Denaturation of DNA.</u> Heat 15 ml of the T4 or T5 DNA solution in a test tube (sealed with aluminum foil) in a boiling water bath for 7 min and immediately pour into an ice-cold 250-ml Erlenmeyer flask containing 1.5 ml 20 × SSC. Thus, the DNA solution is adjusted to about 2 × SSC. Swirl vigorously to cool the DNA solution quickly and to prevent renaturation. Then add 59 ml 2 × SSC, bringing the DNA concentration down to 20 µg/ml.

2. <u>Preparation of the DNA-Nitrocellulose Filter</u> (see data sheet I). Each group prepares 10 filters with DNA from T4 or T5 and 5 control filters without DNA. Soak 15 nitrocellulose filters in 2 × SSC for 5-10 min. Then put 5 of these filters on the sieves of the filter apparatus, using tweezers. Weight down the filters with the stainless steel rings. Be careful to keep original position of the filters. The hole in the stainless steel rings is constructed to take 5 ml. Pass 60 ml 2 × SSC through each filter with gentle suction. Place the filters, face side up into the scintillation vials (control filters Nos. 1-5). Place 10 more filters on the filter box, as described above. Using gentle suction, pull 5 ml 2 × SSC through each filter, and then allow 5 ml of the solution of denatured T4 or T5 DNA to trickle through the filters without suction. Each filter should retain about 100 µg DNA. Using gentle suction again, pull 50 ml of 2 × SSC through each filter and dry the filters on paper towels, at room temperature for about 2 hrs. Place each filter into one scintillation vial, DNA face up, and vacuum dry at 80°C for 4 hrs. The filters can be kept for several weeks in well-closed scintillation vials. These can then be used for further hybridization tests.

3. <u>Hybridization</u> (see data sheet I). Exchange T4 and T5 DNA filters (in scintillation vials) among the groups, so that each group has a set of filters corresponding to data sheet I. For the rest of the experiment, however, only two filters, chosen at random, from each of the three filter groups (control; T4; T5) will be used. Pipette 2 ml of 2 × SSC into each vial and add 20 µl radioactive T4 mRNA and immerse filter completely with the pipette. Incubate vials in a 65°C water bath over night (approx. 15 hrs). Remove filters and wash twice in 50 ml of 2 × SSC in a 100-ml beaker by gently swirling the filter in the SSC. Use fresh SSC for each washing and each filter. Dry the filters in new, open scintillation vials at 80°C in an oven for 20 min. Now add 15 ml toluene scintillation fluid, screw caps on and measure the radioactivity in the scintillation counter for 1-10 min per filter.

Data sheet I

Filter No.	Before drying		Wash ml 2 × SSC
	Filter liquid ml 2 × SSC	5ml DNA from	
1	10	–	50
2	10	–	50
3	10	–	50
4	10	–	50
5	10	–	50
6	5	T4	50
7	5	T4	50
8	5	T4	50
9	5	T4	50
10	5	T4	50
11	5	T5	50
12	5	T5	50
13	5	T5	50
14	5	T5	50
15	5	T5	50

After drying each filter is subjected to the following treatment:

2 ml 2 × SSC ⎫ 20 µl mRNA ⎬ 15 hrs 65°C	HYBRIDIZING
2 × 50 ml 2 × SSC	WASHING
20 min 80°C	DRYING
15 ml toluene scintillant	COUNTING

Evaluation

1. Calculate the percentage of hybridized RNA from the relation-ship between the added radioactivity and the radioactivity re-tained on the filters (data sheet II). It is necessary to have the following data about the T4 mRNA used. This information is to be obtained from the lab assistant or taken from Expt. 8.

$O.D._{260}$	µg RNA/20 µl*	cpm/20 µl
.		

* Conversion of O.D. units to µg RNA is based on the fact that an $O.D._{260}^{1 cm}$ = 1 corresponds to about 42 µg RNA/ml.

Data sheet II: Result of hybridization

Filter	T4 mRNA cpm/filter	% hybridization
Control		
T4 DNA		
T5 DNA		

What does this result prove?

2. Question: In the present experiment about 100 µg DNA were bound to the filter, but less than 20 µg RNA were added. Why would it not be logical to take five times the amount of RNA with lower specific activity?

Literature

BØVRE, K., SZYBALSKI, W.: DNA-RNA Hybridization Techniques. In: Methods in Enzymology, Vol. 21D (eds. L. GROSSMAN, K. MOLDAVE), pp. 350-383. London: Academic Press 1971.

DE LEY, J.: Hybridization of DNA. In: Methods in Microbiology, Vol. 5 A (eds. J.R. NORRIS, D.W. RIBBONS) pp. 311-329. London: Academic Press 1971.

GILLESPIE, D., SPIEGELMAN, S.: A Quantitative Assay for DNA-RNA Hybrids with DNA Immobilized on a Membrane. J. Mol. Biol. 12, 829-842 (1965).

McCARTHY, B.J., CHURCH, R.B.: The Specificity of Molecular Hybridization Reactions. Ann. Rev. Biochem. 39, 131-150 (1970).

MIDGLEY, J.E.M.: Hybridization of Microbial RNA and DNA. In: Methods in Microbiology, Vol. 5 A (eds. J.R. NORRIS, D.W. RIBBONS), 331 p. London: Academic Press 1971.

WARNAAR, S.O., COHEN, J.A.: A Quantitative Assay for DNA-DNA Hybrids Using Membrane Filters. Biophys. Biochem. Res. Communications 24, 554-558 (1966).

Time requirement: 1st day 3 hrs, not counting the time required to dry the filters.
2nd day approx. 2.5 hrs, including the time required for counting the radioactivity.

Appendix Expt. 9

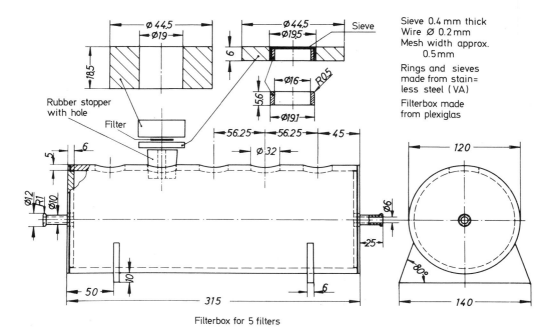

Sieve 0.4 mm thick
Wire Ø 0.2 mm
Mesh width approx.
   0.5 mm

Rings and sieves
made from stain=
less steel (VA)

Filterbox made
from plexiglas

Filterbox for 5 filters

10. Time of Gene Transcription – Determined by Phenocopying with 5–Fluorouracil

During the synthesis of nucleic acids the naturally occuring purines and pyrimidines can be substituted for by halogenated base analogs. The genetic information is then changed because many analogs have pairing properties that differ from those of the native bases. If base analogs are incorporated into re-plicating DNA, e.g., 5-bromodeoxyuridine (Expt. 15), mutations are induced. If base analogs are incorporated into mRNA during transcription, (e.g., 5-fluorouracil (FU)), they induce non-he-reditary "phenocopies". This means that due to exogenous factors an individual shows another phenotype than that which would cor-respond to its own genotype. A mutant, for example, can assume a phenotype similar to that of a wild type.

CHAMPE and BENZER found that in the presence of FU phage mutants which phenocopy the wild type, are predominantly those which arose by a GC $\longrightarrow$ AT transition and which have the mutated purine base in the reading (sense) strand. They interpreted these results with the following model:

Genotype	DNA	mRNA	tRNA	Phenotype
Wild type	C–G X–Y X–Y	C X X	G Y Y $\rfloor$ AA$_1$	Wild type
Mutant without FU	T–A X–Y X–Y	U X X	A Y Y $\rfloor$ AA$_2$	Mutant
Mutant with FU	T–A X–Y X–Y	FU X X	G Y Y $\rfloor$ AA$_1$	Wild type

Abbreviations: AA$_1$, AA$_2$ = Different amino acids.
X–Y = Any pair of complementary bases.

In this example, FU pairs during transcription as a U, and later, during translation, as a C. This molecular mechanism is supported first by the analysis of many mutants with known base pairs at the site of mutation, and second, by the fact that the effect of FU can be neutralized by the addition of an excess of U.

The transcription of phage T4 DNA is a sequential process (com-pare Expt. 8), i.e., different genes or gene clusters are read at different times of intracellular phage development. To decide whether a gene is transcribed "early" or "late" in phage develop-ment, *E. coli* su is infected with the corresponding amber mutant and the cells are then exposed to FU at various intervals during the latent period. Under these circumstances infectious phage progeny (infectious for su$^+$ bacteria) are only formed when the FU is present during transcription of the mutated gene. The addition of FU before or after this moment does not show this effect.

Plan. Non-suppressing (su) *E. coli* bacteria will be infected with an amber mutant of T4 in the presence of FU. After different times of incubation the FU will be diluted out and the number of plaque-forming infected bacteria ("complexes") on a su$^+$ indicator strain will be determined. Estimate from the phage titer at which time the amber mutation is transcribed.

Material. 20 ml of a log culture of *E. coli* BA (su) in NB (30°C, aerated) with a cell titer of 1 × 10^8/ml. 5 ml of a stat culture of *E. coli* CR 63 (su$^+$) in NB (30°C, aerated) as indicator. 0.4 ml of T4am N91 with a titer of 2.5 × 10^9/ml on indicator CR 63. 23 plates with HB agar and 23 HB soft agar tubes, each supplemented with 40 µg/ml uracil. 150 ml of NB. 5 ml of M9 and 100 ml of M9-CAT (= M9 with 0.24% Casamino Acids and 20 µg/ml Tryptophan), both chilled in ice. 2 ml NB containing 0.002 M KCN*. 1 ml each of

- M9-CAT-Thy   : M9-CAT containing 40 µg/ml thymidine
- M9-CAT-Thy-FU   : M9-CAT-Thy containing 20 µg/ml 5-fluorouracil
- M9-CAT-Ura-FU   : M9-CAT supplemented with 40 µg/ml uracil and 20 µg/ml 5-fluorouracil.

The bases and the KCN are directly dissolved in the media, which have already been autoclaved. 1 water bath at 26°C (!). 1 polystyrene pot filled with crushed ice. 2 50-ml centrifuge tubes. For several groups together: 1 SORVALL centrifuge RC2-B with the SS34 rotor. 1 water bath at 37°C. 1 vortex mixer.

Procedure (1st day)

Centrifuge the 20 ml log culture of *E. coli* BA in Rotor SS34 at 6,000 rpm (4,300 × g) for 5 min. Discard supernatant and resuspend the sediment in 1 ml KCN-NB.

1. Phage-adsorption. Mix 0.8 ml of this bacterial suspension and 0.2 ml of T4am N91 and keep in the 37°C water bath for 5 min. Then add 49 ml of ice cold M9-CAT and centrifuge in the SS34 rotor at 6,000 rpm for 10 min. Discard supernatant containing the non-adsorbed phages and KCN, and resuspend the sediment containing the phage-infected bacteria in 2.5 ml ice-cold M9.

2. Phage development**. Pipette 0.5 ml each of the suspension of phage-infected bacteria into test tubes containing

---

* KCN interferes with oxidative phosphorylation. When added during the period of phage adsorption it prevents unsynchronized phage development.
** Each group should perform test (b) plus either (a) or (c).

a) 0.5 ml M9-CAT-Thy   : Control I

b) 0.5 ml M9-CAT-Thy-FU   : Test

c) 0.5 ml M9-CAT-Ura-FU   : Control II (alternative to I)

and immediately incubate in a water bath at 26°C. Following
the data sheet, remove 0.1 ml samples at different times, dilute
in NB and plate with *E. coli* CR63 as indicator. Use only HB-agar
and HB-soft agar containing uracil, in order to replace the remaining FU competitively. Incubate plates at 30°C or 37°C for
about 18 hrs.

Evaluation (2nd day)

Count the plaques on the plates, calculate titers and plot
semi-logarithmically against the time at which the sample was
taken. At which time after infection is the mutated gene in
T4am N91 transcribed at 26°C?

Literature

CHAMPE, S.P., BENZER, S.: Reversal of Mutant Phenotypes by
   5-Fluorouracil: An Approach to Nucleotide Sequences in
   Messenger-RNA. Proc. Nat. Acad. Sci. US 48, 532-546 (1962).

EDLIN, G.: Gene Regulation during Bacteriophage T4 Development I.
   Phenotypic Reversion of T4 Amber Mutants by 5-Fluorouracil.
   J. Molec. Biol. 12, 363-374 (1965).

MANDEL, H.G.: The Incorporation of Fluorouracil into RNA and
   its Molecular Consequences. In: Progress in Molecular and
   Subcellular Biology (ed. F.E. HAHN) 1, 82-135 (1969).

Time requirement: 1st day 2hrs, 2nd day 2 hrs.

Data sheet

Time min	Dilution	Plate No. (a)	(b)	Plaques	Titer
0	10^{-2}	1	–		
		2	–		
	10^{-2}	–	3		
		–	4		
5	10^{-2}	5	–		
		6	–		
	10^{-2}	–	7		
		–	8		
10	10^{-2}	9	–		
		10	–		
	10^{-2}	–	11		
		–	12		
20	10^{-2}	13	–		
		14	–		
	10^{-2}	–	15		
	10^{-3}	–	16		
		–	17		
30	10^{-2}	18	–		
		19	–		
	10^{-2}	–	20		
	10^{-3}	–	21		
		–	22		
	10^{-4}	–	23		

a) T4 am-infected bacteria in M9-CAT-Thy or alternatively in M9-CAT-Ura-FU.
b) T4 am-infected bacteria in M9-CAT-Thy-FU.

Problems

1. DNA consists of the following four nucleotides:

Deoxyadenosine monophosphate (dAMP) $C_{10}H_{14}O_6N_5P$

Deoxythymidine monophosphate (dTMP) $C_{10}H_{15}O_8N_2P$

Deoxyguanosine monophosphate (dGMP) $C_{10}H_{14}O_7N_5P$

Deoxycytidine monophosphate (dCMP) $C_9H_{14}O_7N_3P$

The molecular weights of these nucleotides are to be calculated and averaged.
Note: In the English literature molecular weights are given in "Dalton" units; one dalton corresponds to an atomic weight of 1.

2. Double-stranded DNA of phage T4 has a molecular weight of 130×10^6 daltons. What is the mass (in g) of such a DNA molecule?
Note: AVOGADRO's Number = 6.02×10^{23} molecules per mole.

3. The small phage ΦX174 (Phi X) contains single-stranded DNA with a molecular weight of 1.7×10^6 daltons.
a) How many nucleotides correspond to this molecular weight?
b) How many different proteins can be coded for by this amount of DNA, if the estimation is based on the triplet code (3 nucleotides $\triangleq$ 1 amino acid) and on the assumption that proteins consist of 250 amino acids, on the average?

4. When double-stranded DNA is carefully extracted from Lambda phages, it has an average length of 17.2 µm, as estimated from electron micrographs. What molecular weight corresponds to this length, if a DNA-segment of 10 nucleotide pairs has a length of 34 Å units?
Note: Use the average molecular weight calculated in Problem 1. When nucleotides are polymerized, 1 molecule of H_2O is released per ester bond; however, the relationship "34 Å per 10 nucleotide pairs" is valid only for hydrated DNA (= B-configuration), in which 1 molecule of H_2O on the average is bound to a nucleotide (1 mm = 10^3 µm= 10^6 nm = 10^7 Ångström units).

5. Suppose that DNA consists of equimolar quantities of dAMP, dTMP, dGMP and dCMP. What percent increase in molecular weight is gained by the following substitutions?

a) all ^{14}N atoms by ^{15}N?

b) all ^{12}C atoms by ^{13}C?

c) all ^1H atoms by ^2H?

d) all dTMP by dBUMP (deoxybromouridine monophosphate)?
Note to (d): If the $-CH_3$ group in dTMP is substituted by $-Br$ (bromine), dBUMP is obtained.

6. Suppose a DNA molecule is composed statistically of the usual 4 bases, and consists of 3.5×10^6 base pairs. What would

be the minimum chain length of any arbitrary sequence if it were to appear only once according to statistical expectation in the above mentioned DNA molecule?

Note: Base sequences starting at a certain minimal length occur only once in a given DNA molecule. This is probably a basic requirement for the specific binding of repressor molecules to their respective operators, the proper pairing of homologous DNA double strands during recombination, and other processes which need recognition sites on DNA.

7. How many base pairs are to be expected in the genes for
a) the β-polypeptide of the DNA-dependent RNA polymerase of *E. coli*: MW approx. 155,000, corresponding to about 1,400 amino acids.
b) Phage T4 lysozyme consisting of 164 amino acids.
c) Yeast tRNA $_{Ala}$, consisting of 77 nucleotides?

C. Mutation and Photobiology

Mutations are defined as heritable changes in the genetic material. They generally occur as rare and undirected events. The mutability of DNA is a basic requirement for the phylogenetic development of organisms because only mutations create that variety of different genotypes which allows other evolutionary factors, e.g. selection and recombination, to act. Much of today's knowledge of classical and molecular genetics as well as of biochemical physiology is based on studies with mutants which are blocked in single gene functions.

Mutations may be classified as follows:

a) Genome-mutations: Heritable changes in the number of single chromosomes or of sets of chromosomes per nucleus.

b) Chromosome-mutations: Heritable changes in the structure of chromosomes by loss, duplication or rearrangement of chromosomal segments, which often comprise more than a single gene.

c) Gene or point-mutations: Heritable changes of single or several base pairs. A distinction is made between base pair exchanges (transitions; transversions), and changes in the number of base pairs per gene (insertions; deletions).

Mutations of type b and c are frequently observed in prokaryotic microorganisms and viruses, which usually contain only a single DNA molecule per cell or virus particle. In a formal sense, even genome mutations occur among some bacteria since it is possible to isolate mutants with a heritable change in the number of certain chromosomes (plasmid DNA molecules) per cell. The molecular mechanism of mutations is best understood in the case of the point mutations. This is due to the fact that they are most easily inducible and that many of the chemical substances used as mutagens act highly specifically on the base-sequence of nucleic acids.

Gene mutations are usually multi step events: they are initiated by "premutations" which then require one or more DNA replications in order to replace the original base pair at the mutation site by another. Examples of premutations are the deamination of adenine to hypoxanthine by nitrous acid or the substitution of cytosine by 5-bromouracil (BU) in replicating DNA. Since hypoxanthine and 5-bromouracil show base-pairing characteristics that differ from those of bases which were originally at the mutation site "mispairing" will occur during DNA replication. The reaction steps finally leading to the mutation in these two examples are shown schematically as follows.

A	+ HNO$_2$	A*	Replic.	A*	Replic.	G	TRANSITION
T	$\longrightarrow$	T	$\longrightarrow$	C	$\longrightarrow$	C	

G	+ BU	G	Replic.	A	Replic.	A	TRANSITION
C	$\longrightarrow$	BU	$\longrightarrow$	BU	$\longrightarrow$	T	

(A* stands for hypoxanthine. The nucleotide not deaminated or substituted for has been omitted from the scheme.)

Viruses with single-stranded RNA can mutate without passing through a premutational state, if cytidine is deaminated to uridine by nitrous acid.

Generally a mutational change of a given base sequence (forward mutation) can only be reverted by another mutation (back mutation). Some premutational DNA changes, however, can be eliminated enzymatically. Thus for instance UV-induced pyrimidine dimers can be split into monomers by a photoreactivating enzyme before they give rise to a mutation.

Apart from the hypothesis of "mutation by mispairing" which has been described, there exist others which have been experimentally less well investigated, such as the hypothesis of mutations caused by faulty repair of DNA.

Many investigations of mutations begin with methods designed to distinguish between mutation and modification. An example of this will be given in Expt. 11. Spontaneously, chemically and physically induced mutations are dealt with in Expts. 12, 13 and 15. Expts. 12 and 13 also show the calculation of mutation rates and some statistical analyses (χ^2-test) of the experimental data. The mutagen test with filter disks (Expt. 15) is evaluable only semiquantitatively. However, it demonstrates a simple and frequently used screening technique to detect chemical mutagens. Finally Expt. 14 is thought to be representative of many *in vitro* tests with mutagens and precursors of DNA or RNA all of which aim to describe mutations in terms of defined chemical reactions.

Literature

DRAKE, J.W.: The Molecular Basis of Mutation, 273 pp. London: Holden-Day 1970.

HOLLAENDER, A. (ed.): Chemical Mutagens. Principles and Methods for their Detection, Vol. I and II. London: Plenum Press 1971.

SMITH, K.C., HANAWALT, P.C.: Molecular Photobiology-Inactivation and Recovery, 230 pp. London: Academic Press 1969.

11. Genotype and Phenotype

Each living cell and each virus is characterized by its genotype and its phenotype. The genotype is equivalent to the sum
of all the genes and thus represents the potential abilities of
an individual. The phenotype results from both the genotype and
the respective environment in which the hereditary information
is expressed. Until now, conclusions as to the genotype of an
individual could only be drawn by inference from its phenotype
as expressed under different environmental conditions. If there
is a phenotypic difference between two individuals belonging to
the same species and having developed in the same environment,
and furthermore if this difference is maintained by their vegetative progeny, then this difference is considered to be hereditarily determined. Two different genotypes are represented by,
for example, a "wild type" and a "mutant". Note, however, that
if both individuals and their vegetative progeny show the same
phenotype, they can still have different genotypes. Many genotypic differences cannot be detected or can be detected only
under special environmental conditions. Changes in the genotypes
can occur by mutations which change the structure or the number
of the chromosomes. Changes of phenotype which are not based on
mutations are called modifications and are caused by variation
of the chemical or physical environment, i.e., the nutrition or
the temperature.

A bacterial colony (clone) consists of the vegetative and, therefore, in general isogenic progeny of a single bacterium. If a
colony is grown under defined cultural conditions its appearance
allows to draw conclusions as to the phenotype and the genotype
of the individual cells (approx. 10^7-10^8). When all the colonies
of one agar plate are to be transferred to other agar plates containing different media in order to analyze the progeny under
different growth conditions, it is not necessary to make monotonous single transfers with the inoculating loop. Using a sterile
velvet stamp, the entire colony pattern can be transferred at
once (LEDERBERG Replica Plating Technique).

Plan. There are 50-100 bacterial colonies of the identical phenotypical appearance on nutrient broth agar plates. From these colonies, "subcultures" are to be started under varying environmental conditions (different nutrient media and temperatures), in
order to determine which genotypes and how many colonies of each
exist on the original agar plates. Using a colony characteristic
whose expression is especially dependent on temperature, the effect of temperature will be discussed.

Material. 4 plates with NB agar ("master plates") each with
50-100 evenly distributed bacterial colonies, which have grown
at 38°C (!) during a two-day incubation period. Each plate was
seeded with cells of a mixture of different strains (wild-type
and mutants) of *Serratia marcescens* (for details see Chapter IV/A).
4 sterile plates with NB agar and 4 with M9 agar. 1 round wooden
block "stamp" (diameter 7.9 cm; height approx. 7 cm) with a rub-

ber band; 4 sterile velvet pads (approx. 16 × 16 cm), wrapped
in paper (they were autoclaved at 1 atm for 15 min and then dried
at 30°C for 4 days).

Procedure

First day: Check the colonies on the master plates: what appearance do they have? Do they all look alike? Note eventual deviations
on the data sheet and mark with a crayon on the bottom of the plate.
Number the M9 agar plates on the bottom from 1 to 4 and the NB-
agar plates correspondingly from 5 to 8; the numbers serve as an
orientation aid for stamping and comparing the plates. Tightly
stretch the velvet cloth over the wooden block; touch the velvet
on its edge only. Transfer the colony pattern of a master plate
onto the velvet as shown in the illustration. Carefully transfer ("replicate") the colony pattern, which is now on the velvet,
first to the M9 agar plate 1 and then to the NB agar plate 5. It
is important to treat the plates in this order. Using new sterile
velvet, repeat the same procedure, with plates 2 + 6, 3 + 7 and
4 + 8, respectively. Incubate the replica plates at 30°C for 1
day and the master plates at 38°C one day longer.

I   II   III

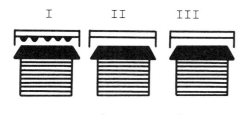

I Master plate

II M9 agar plate

III NB agar plate

Stamp: Lined

Second day: Count the colonies on all 8 replica plates and
compare their colony pattern with those of the corresponding
master plates; count colonies of different pigmentation separately (data sheet I). How many and which genotypes were on
the master plates? What is their frequency (data sheet II)?
Other observations?

An explanation is required as to why certain colonies are red
when incubated at 30°C and colorless at 38°C. Which of following
alternative hypotheses is probably correct and why? The red pigment is not formed at 38°C because

- the mRNA necessary for its synthesis is thermolabile (Hypothesis a)
- the pigment itself is thermolabile (Hypothesis b)
- at least one enzyme which is involved in the synthesis of
  the pigment is thermolabile (Hypothesis c).

What simple test would exclude one of the alternatives?

Why is the colony pattern of the original plate first stamped
on M9 agar and then on NB agar?

Data sheet I

Plate No.		Number of colonies	
		red (pig$^+$)	colorless (pig)
M9/1 M9/2 M9/3 M9/4			
	Sum:	S_I =	S_{II} =
NB/5 NB/6 NB/7 NB/8			
	Sum:	S_{III} =	S_{IV} =

Data sheet II

Frequency (%)	Genotype (mark correct answer)
100 x $S_I/(S_{III} + S_{IV})$ = ...	prototr., auxotr., pig$^+$, pig
100 x $S_{II}/(S_{III} + S_{IV})$ = ...	prototr., auxotr., pig$^+$, pig
100 x $(S_{III} - S_I)/(S_{III} + S_{IV})$ = ...	prototr., auxotr., pig$^+$, pig
100 x $(S_{IV} - S_{II})/(S_{III} + S_{IV})$ = ...	prototr., auxotr., pig$^+$, pig
Sum	

Literature

LYNCH, D.E., WORTHY, T.E., KRESCHECK, G.C.: Chromatographic Separation of the Pigment Fractions from a *Serratia Marcescens* Strain. Applied Microbiology 16, 13-3O (1968).

RAPOPORT, H., HOLDEN, K.G.: The Synthesis of Prodigiosin. J. American Chemical Society 84, 635-642 (1962).

WILLIAMS, R.P., GOLDSCHMIDT, M.E., GOTT, C.L.: Inhibition by Temperature of the Terminal Step in Biosynthesis of Prodigiosin. Biochem. Biophys. Res. Com. 19, 177-181 (1965).

Time requirement: 1st day 0.5 hrs, 2nd day 2 hrs.

12. Proof of the Random Distribution of Spontaneous Mutations

Spontaneous mutations are usually triggered by endogenous factors,
e.g., a defective DNA-polymerase or an unequal crossover. Since
spontaneous mutations occur rarely their appearance is only quantitatively conceivable in large populations. The frequency of mutations follows statistical laws. Thus it is important to know
whether spontaneous mutations in cells occur at random and independently of one another. This question can be answered with
the papilla test: approx. 100 cells of a histidine requiring
(his) mutant of *E. coli* will be spread on minimal agar which contains only traces of histidine. The cells divide and form small
colonies as long as there is sufficient histidine available.
During the development of colonies, spontaneous mutations occur;
among these are mutations to histidine independence. These his^+
cells have a growth advantage as compared to the his cells, since
they no longer require the limited histidine in the agar. They
grow as a clone and form papillae or microcolonies which are then
found on top of the colonies of the his cells. Each papilla represents one single mutation to prototrophy. If the mutations
his $\longrightarrow$ his^+ occur at random during the development of the his
colonies, i.e., if each cell has the same small chance to mutate
per time unit, the numerical distribution of the papillae on the
colonies must fit a POISSON distribution. In contrast to the
GAUSSIAN distribution, the POISSON distribution is asymmetrical.
It is asymmetrical because it is valid only for rare "events".
If, for example, an average of 2 papillae per colony are found,
the deviations are unilaterally limited; namely, less than zero
papillae per colony are not possible, but 3 and 4 and even more
papillae per colony are found. The POISSON formula in its general
form states:

$$P_n = \frac{m^n \times e^{-m}}{n!}. \tag{1}$$

This means that "events" (e.g., mutations) will be found in
random samples with a probability of P_n, if the "expected value"
(which is "the average frequency obtained from many random samples") is m. e is the natural base of logarithms (= 2.718). As
n is a whole number variable (0, 1, 2, 3 ...), the following
special terms can be obtained from formula (1):

$$P_0 = e^{-m} \tag{2}$$

$$P_1 = m \times e^{-m} = m \times P_0 \tag{3}$$

$$P_2 = m^2 \times e^{-m}/2! = \frac{m}{2} \times P_1 \tag{4}$$

$$P_3 = m^3 \times e^{-m}/3! = \frac{m}{3} \times P_2 \tag{5}$$

$$P_{\geqslant 1} = 1 - P_0 = 1 - e^{-m} \tag{6}$$

$$P_{\geqslant 2} = 1 - P_0 - P_1 = 1 - e^{-m}(1 + m) \tag{7}$$

In the papilla test

$$m = \frac{\text{Sum of all the papillae}}{\text{Sum of all the colonies}}$$

and P_n indicates which frequency, according to POISSON, is to be expected for colonies with n = O or 1 or 2 or 3, etc. papillae.

Plan. Growing colonies of a histidine-requiring mutant of *E. coli*, grown on minimal agar containing limiting histidine, will be counted, and will be classified as colonies having either O, 1, 2, 3, or ≥4 papillae. Each papilla corresponds to a mutation to prototrophy during the development of the colony. The experimentally determined distribution of the papillae on the colonies will be examined with the Chi Square test, to verify that it fits a POISSON distribution. The cells of several colonies (3 with papillae and 3 without papillae) will be streaked onto minimal agar plates for genotype analysis. The spontaneous mutation rate of his to his$^+$ will be estimated.

Material. 4 plates with histidine-supplemented minimal agar, according to RYAN, and colonies of *E. coli* 15 his with and without papillae (approx. 1OO colonies per plate, grown at 37°C for 2-3 days). 1 dissecting microscope (magn. 5-10 times). 2 minimal (M9) agar plates.

Procedure

1. Divide two minimal agar (M9) plates on their bottom sides into 3 sectors with crayon. Select 3 colonies free of any papilla with the dissecting microscope and streak a large sample of each onto one of the three sectors of an M9 agar plate. Repeat the same procedure with samples from papillae of 3 different colonies and the second M9 agar plate.

M9 agar plate No.	Growth in Sector		
	I	II	III
1			
2			

2. Count all colonies of approximately equal size and list separately those with 1, 2, 3 and ≥4 papillae. Do not consider very large or very small colonies because this would disturb the POISSON analysis. Colonies used under (1) must be included in the analysis.

3. Take five equally large papilla-free colonies in their entirety, i.e., with the agar underneath. Suspend them together in 1 ml P-buffer, and determine the cell titer microscopically in a counting chamber. Convert cell titers into number of cells (N) per colony.

Evaluation

1. Determine the Chi-square and the level of confidence P (see data sheet). If the distribution of the papillae on the colonies does not correspond to the POISSON distribution, discuss possible reasons for this.

2. Calculate the spontaneous rate, α, for the mutation of his to his[+] according to RYAN and co-workers (1955):

$$\alpha = m \times \ln 2/N = \ldots\ldots\ldots \text{ (per cell and generation).}$$

This value is smaller than the value given in the literature (maximally by about a factor of 2), because some of the papilla-free colonies do contain back-mutants (see RYAN et al., 1955). How can this be proven?

Literature

KLECZKOWSKI, A.: Experimental Design and Statistical Methods of Assay. In: Methods in Virology 4, 616-730. Esp. Section IV (Frequency Distributions). London: Academic Press 1968.

LEA, E.E., COULSON, C.A.: The Distribution of the Numbers of Mutants in Bacterial Populations. J. Genetics 49, 264-285 (1949)*.

LURIA, S.E., DELBRÜCK, M.: Mutations of Bacteria from Virus Sensitivity to Virus Resistance. Genetics 28, 491-511 (1943)*.

RYAN, F.J., SCHWARTZ, M., FRIED, P.: The Direct Enumeration of Spontaneous and Induced Mutations in Bacteria. J. Bacteriol. 69, 552-557 (1955).

Time requirement: 1st day 2 hrs, 2nd day 15 min.

---

* Republished in LEDERBERG, J. (ed.): Papers in Microbial Genetics, 303 p. Madison: University of Wisconsin Press 1952.

Data sheet

	Experimental results			Theoretical values (POISSON)		Chi square		
Papillae (n) per colony	Colonies (Q$_n$) with n papillae	Sum of the papillae (n × Q$_n$)	Probability (P$_n$) of colonies with n papillae	Expected number (E$_n$) E$_n$ = P$_n$ × ΣQ$_n$	Difference Δ_n = Q$_n$ − E$_n$	Δ_n^2	$\dfrac{\Delta_n^2}{E_n}$
0							
1							
2							
3							
≥ 4							
	Σ =	Σ =	Σ = 1.00	Σ =	Σ = 0	./.	Σ =

$$\chi^2 =$$

Average frequency of papillae m $= \dfrac{\Sigma n \times Q_n}{\Sigma Q_n} = \qquad =$

The level of confidence P = (see Figs. 4, 5 and 6)

The difference between result and random distribution is

P ≤ 0.01 significant

P : 0.01 – 0.05 probable

P > 0.05 uncertain

13. Mutagenic and Inactivating Action of Ultraviolet Light on Phage Kappa

Short-wave ultraviolet light (UV) of about 254 nm inactivates and mutates phages. A phage is considered to be "inactivated" or lethally damaged when it no longer propagates, i.e., when it can no longer form a plaque. A phage is "mutated" when its progeny differs genetically from the parental phage. Kappa is a temperate phage and therefore usually forms turbid plaques. Spontaneous or induced mutations can lead to clear plaques. In comparison to the wild type, these so-called c (clear) mutants cannot stably lysogenize the host cells.

If phages are irradiated for different times and the frequency of the survivors and of the mutants among the surviving phages is plotted as a function of time, "dose-effect curves" are obtained. These curves correspond to a one, two or more "hit" reaction. Conclusions may, therefore, be drawn as to the molecular processes of inactivation and mutation. The term "hit" is used, because radiation energy in the form of single quanta hits the molecules. "One-hit" kinetic means that a single hit is sufficient to induce a mutation. When phages are irradiated with short-wave UV light many quanta are absorbed in the nucleic acids without inducing any photochemical changes in the bases. Furthermore, even if an "effective" absorption of an UV quantum has taken place, i.e., if the DNA at a certain site has been photochemically changed, this need not result in an inactivation or a mutation. As soon as the damaged phage has injected its nucleic acid into the host cell, one or more enzymes can repair the initiated UV-damages. One such a "repair mechanism" is Host Cell Reactivation or HCR, which is genetically controlled by the host bacterium. If bacteria lose the capacity for HCR, because of a mutation in one of their hcr genes, they become highly UV-sensitive. If an hcr mutant is used as a host for UV-irradiated temperate phages, these have a much lower chance to survive than when plated with the wild-type indicator strain. HCR requires double-stranded DNA. It occurs in only one or the other of the two DNA strands for any given damaged segment. The most important characteristic of HCR is that a specific endonuclease excises from the DNA those segments which contain photochemically altered nucleotides, especially pyrimidine dimers. The gaps which are formed by this process are closed by "repair replication" in which DNA is synthesized using the strand opposite the gap as a template.

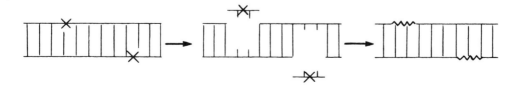

Plan. A suspension of kappa phages will be irradiated with UV light. Samples will be removed at different times of irradiation and will be plated with *Serratia marcescens* HY as indicator. The frequency of survivors (N/N_0) and the frequency of clear-plaque mutants among the survivors will be plotted as a function of the irradiation time ("dose-effect curve"). It will be determined whether the induction of the mutations follows one- or two-hit kinetics. In addition to the wild type (hcr^+) an hcr mutant will be used as an indicator for phages irradiated for 0 and 1 min, in order to estimate the extent of host-cell reactivation.

Material. 10 ml of a suspension of phage kappa, wild type (6×10^7/ml) in P-buffer. 10 ml of a stat culture in NB of *Serratia marcescens* W225 (hcr^+) and W227 (hcr) as indicator (cell titer approx. 1×10^9/ml). 1 sterile petri dish (Ø 9.0 cm) with a magnetic rod for stirring. 34 plates with NB agar. 34 NB soft agar tubes. 1 yellow light lamp. 1 sheet of graph-paper. Nomograms for standard deviations and the Chi square test. For several groups together: 1 ozone-free Hg low pressure (ultraviolet) lamp in a metal housing with a sliding shutter. Lamps of this type mainly emit radiation of the Hg-resonance line 253.7 nm. 1 magnetic stirrer. The distance between the bottom side of the lamp and the top side of the magnetic stirrer depends on the energy output of the lamps. Using the UV lamp OSRAM HNS 12 the right distance would be about 36 cm and the energy flux rate ("intensity" = photon-flux rate) at the level of the phage suspension to be irradiated is then about 11 ergs mm^{-2} sec^{-1}*.

Procedure

1. Prepare all dilution tubes with P-buffer and number 34 NB agar plates (see data sheet I). Turn on the UV lamp 15 min before starting the irradiation of phages. Keep the laboratory dark and work in dim yellow light, in order to avoid undesired photoreactivation. Dilute a 0.1 ml aliquot of the non-irradiated phage suspension and prepare plates with the agar layer technique, using W225 (plates Nos. 1-6) and W227 (plates Nos. 31 and 34) as indicators.

2. Irradiate the rest of the phage suspension in a sterile petri dish with UV (using protective goggles!) and stir magnetically. After 1, 2 and 3 min interrupt the irradiation by closing the shutter of the lamp (do not switch off UV light!), take a sample and immediately continue irradiation. Dilute samples and plate (see date sheet I). Incubate plates at 30°C for about 18 hrs.

---

* These approximate values were obtained with the aid of T2 phages and are based on the fact that an energy flux of about 20 erg mm^{-2} is needed for UV inactivation (254 nm) of T2 to a survival of 37% (= 1 lethal hit).

Evaluation (2nd day)

1. Screen all plates carefully for clear plaques ("mutants") using the magnifying glass (see data sheet I). Afterwards count all plaques ("total, N_i") on plates 4-6, 10-12, 16-18, 22-24, 28-30, and 31-34 and calculate the phage titers. Estimate the number of plaques on the rest of the plates using these titers.

2. Phage-inactivation. Calculate the log N/N_0 values from the phage titers which were obtained with W225 (hcr$^+$) as indicator, and plot on graph-paper as a function of the irradiation time, t. If all points fit closely to a straight line which runs through the origin, ex- ponential inactivation (of single hit) is spoken of, because this straight line corresponds to the exponential equation $N/N_0 = e^{-kt}$. Here -k means the inactivation constant which is to be calculated according to:

$$-k_{W225} = \frac{2.30 \ (\log N - \log N_0)}{t} = \underline{\hspace{3cm}} = \underline{\underline{\hspace{2cm}}} \ (\min^{-1})$$

-k gives the slope of the semi-logarithmic inactivation curve (compare growth constant k in Appendix to Expt. 1). For the calculation of -k, any N which lies exactly on the straight line in the graph might be inserted.

Assuming that, with W227 hcr as indicator, the inactivation of kappa is also exponential, we can calculate

$$-k_{W227} = \frac{2.30 \ (\log N - \log N_0)}{t} = \underline{\hspace{3cm}} = \underline{\underline{\hspace{2cm}}} \ (\min^{-1})$$

3. The fraction of inactivating damages sensitive to HCR. Calculate the ratio

$$\frac{(k_{W227} - k_{W225}) \times 100}{k_{W227}} = \underline{\hspace{3cm}} = \underline{\underline{\hspace{2cm}}} \ \%$$

This states what fraction of the original inactivation damages caused in the phages was removed by host-reactivation in the W225 bacteria. The "reactivating sector" can also be graphically found according to $1 - t/t_r$. Here t or t_r means the irradiation time by which the same level of survival of the phages on W227 (hcr) or W225 (hcr$^+$) is achieved.

4. The absolute radiation sensitivity of the phages kappa and T2. The zero term of the POISSON formula ($P_0 = e^{-1} = 0.368$) indicates that all phages of an irradiated suspension received a single lethal hit on the average, when 36.8% survive (presuming the inactivation is negatively exponential). For which time (t)

does kappa have to be irradiated, in order to reach a survival of 37% on W225 (hcr$^+$) and W227 (hcr)?

$$t_{W225} = \frac{2.30\,(logN - logN_0)}{k_{W225}} = \underline{\hspace{3cm}} = \underline{\underline{\hspace{3cm}}} \quad \text{(min)}$$

$$t_{W227} = \frac{2.30\,(logN - logN_0)}{k_{W227}} = \underline{\hspace{3cm}} = \underline{\underline{\hspace{3cm}}} \quad \text{(min)}$$

Here N_0 = 100 and N = 37.

Convert both of these t-values into energy flux ("dose D") values. Our irradiation conditions were 11 ergs mm^{-2} sec^{-1}.

$D_{W225} = \underline{\hspace{4cm}}$ ergs mm^{-2}

$D_{W227} = \underline{\hspace{4cm}}$ ergs mm^{-2}

The corresponding dose value for phages T2 whose DNA content is about 2.5 times higher than that of kappa, amounts to 20 ergs mm^{-2}. What conclusions can be drawn from a comparison of the three dose values?

5. Induction of mutations*. Calculate for each irradiation time (t) the frequency (h_i) of the clear-plaque mutants (see data sheet II):

$$h_i = \frac{M_i}{N_i} = \frac{\text{Mutant plaques}}{\text{Total number of plaques}}$$

Plot the h_i values together with the standard deviation (s_i) as a function of t (graph-paper). Take s_i values from the nomogram for the standard deviation on p. 12.

Calculate the mutation rate μ, i.e., the probability of a mutation per unit of dose. Use data sheet II and the Appendix to Expt. 13.

For one-hit mutation kinetics

$$\mu = \Sigma M_i / \Sigma N_i t_i = \underline{\underline{\hspace{4cm}}} \quad (min^{-1}).$$

The corresponding theoretical dose curve for the relative mutation frequency is then given by

$$p_i = 1 - e^{-\mu t}$$

$$p_i \approx \mu t$$

---

* The UV-induced increase in the frequency of mutation is not due to selection of spontaneous mutants in this case. This was proven in "reconstruction tests" with defined mixtures of wild-type and mutants. Furthermore, the frequency of spontaneous mutants is assumed to be negligibly small.

For <u>two-hit</u> mutation kinetics

$$\mu^2 = \Sigma M_i / \Sigma N_i t_i^2 = \underline{\hspace{4cm}} \quad (min^{-2})$$

and the corresponding theoretical dose curve is given by

$$p_i = 1 - e^{-\mu t} - \mu t \times e^{-\mu t}$$

$$p_i \approx \mu^2 t^2$$

Check the difference between experimental (h-values) and one- or two-hit hypotheses (p-values) for significance with the <u>Chi-square test.</u>

$$\text{Chi-square} = \chi^2_{n-1} = \sum_{i=1}^{n} \left(\frac{h_i - p_i}{s_i} \right)^2$$

(n-1) Number of degrees of freedom.

To obtain the P-values for each χ^2 use the tables on pp. 15-17.

The deviation between the experimental result and the theoretical expectation (one hit or two hit process) is

significant if $P \leqslant 0.01$

probable   if $P = 0.01 - 0.05$

uncertain   if $P > 0.05$

<u>Literature</u>

BRIDGES, B.A.: Mechanism of Radiation Mutagenesis in Cellular and Subcellular Systems. Annual Review of Nuclear Science 19, 139-178 (1969).

EISENSTARK, A.: Mutagenic and Lethal Effects of Visible and Near-Ultraviolet Light on Bacterial Cells. In: Advan. Genetics 16, 167-198 (1971).

KLECZKOWSKI, A.: Methods of Inactivation by Ultraviolet Radiation. In: Methods in Virology 4, 93-138 (1968).

WINKLER, U.: Wirtsreaktivierung von extrazellulär strahlen-induzierten Prämutationen im *Serratia*-Phagen Kappa. Zeitschrift f. Vererbungslehre 97, 18-39 (1965).

WITKIN, E.M.: Ultraviolet-Induced Mutation and DNA Repair. Annual Review of Genetics 3, 525-552 (1969).

<u>Time requirement:</u> 1st day 3 hrs, 2nd day 6 hrs.

Data sheet I

UV t_i (min)	Dilution	Plate No.	No. of plaques mutants M_i	total N_i	Plaque titer	log N	$\log\frac{N}{N_0}$
0	1×10^{-4}	1					
		2					
		3					
	3×10^{-5}	4					
		5					
		6					
		Sum:			–	–	–
1	5×10^{-4}	7					
		8					
		9					
	1×10^{-4}	10					
		11					
		12					
		Sum:			–	–	–
2	1×10^{-3}	13					
		14					
		15					
	3×10^{-4}	16					
		17					
		18					
		Sum:			–	–	–

UV t_i (min)	Dilution	Plate No.	No. of plaques mutants M_i	total N_i	Plaque titer	log N	$\log\frac{N}{N_0}$
3	5×10^{-3}	19					
		20					
		21					
	1×10^{-3}	22					
		23					
		24					
		Sum:			–	–	–
4	1×10^{-2}	25					
		26					
		27					
	3×10^{-3}	28					
		29					
		30					
		Sum:			–	–	–
0	3×10^{-5}	31					
		32					
1	undiluted	33					
		34					

Nos. 1-30: Indicator W225; Nos. 31-34: Indicator W227

Data sheet II

One-hit hypothesis

t_i	N_i	M_i	$h_i \pm s_i$ (%)	$N_i t_i$	$p_i = \mu t_i$ (%)	$D_i = h_i - p_i$ (%)	$\dfrac{D_i}{s_i}$	$\left(\dfrac{D_i}{s_i}\right)^2$
1								
2								
3								
4								
	Σ			Σ				$\chi_3^2 =$

Two-hit hypothesis

t_i^2	N_i	M_i	$h_i \pm s_i$ (%)	$N_i t_i^2$	$p_i = \mu^2 t_i^2$ (%)	$D_i = h_i - p_i$ (%)	$\dfrac{D_i}{s_i}$	$\left(\dfrac{D_i}{s_i}\right)^2$
1								
4		values						
9		see above						
16								
				Σ				$\chi_3^2 =$

P = _____ for one-hit hypothesis; P = _____ for two-hit hypothesis

Appendix to 13: Derivation of Theoretical Dose Curves

a) ONE-HIT Mutation kinetics. According to POISSON (compare Expt. 12) the probability for non hit phages is

$$P_0 = e^{-m}. \tag{1}$$

Here, $m = \mu t$ or the average number of mutation hits per phage and irradiation time t ("dose"). The probability for phages with at least <u>one</u> mutation hit would then be

$$P_{\geqslant 1} = 1 - e^{-m} = 1 - e^{-\mu t}. \tag{2}$$

Correspondingly, the theoretically expected frequency of phages with at least one mutation hit for any given t is

$$P_I = 1 - e^{-\mu t}. \tag{3}$$

If the following progression is developed:

$$e^{-\mu t} = 1 - \frac{\mu t}{1!} + \frac{(\mu t)^2}{2!} - \frac{(\mu t)^3}{3!} + \ldots$$

then Eq. (4) is valid for $\mu t < 0.1$ and correspondingly p<10%

$$e^{-\mu t} \approx 1 - \mu t. \tag{4}$$

By insertion in Eq. (3), one gets

$$P_I \approx 1 - (1 - \mu t)$$

$$\underline{P_I \approx \mu t.} \tag{5}$$

If spontaneous mutants with the frequency p are already present in the phage population before treatment with a mutagen, mutations can only be induced in the fraction $1 - p_0$ of the total population. From Eq. (3), we then get

$$P_I = 1 - (1 - p_0) \times e^{-\mu t} \tag{6}$$

and by making an allowance for (4)

$$\underline{P_I \approx p_0 + \mu t.} \tag{7}$$

b) TWO-HIT Mutation kinetics. According to POISSON the probability for phages with at least two mutation hits is

$$P_{\geqslant 2} = 1 - e^{-m} - m \times e^{-m}. \tag{8}$$

At any given irradiation time, t, the theoretical frequency of phages with at least 2 mutation hits is then

$$P_{II} = 1 - e^{-\mu t} - \mu t \times e^{-\mu t} \tag{9a}$$

$$P_{II} = 1 - e^{-\mu t} (1 + \mu t). \tag{9b}$$

If the approximation (4) is taken into account, we obtain

$$P_{II} \approx 1 - (1 - \mu t) \times (1 + \mu t)$$

$$\underline{P_{II} \approx (\mu t)^2}. \tag{10}$$

If spontaneous mutants are present, the Eq. (11) is analogous to approximation (7)

$$\underline{P_{II} \approx p_0 + \mu^2 t^2}. \tag{11}$$

c) The value μ, which is necessary for the calculation of the theoretical dose curves, is obtained from the mutant frequency (h) found in the experiment, according to the Method of Maximum Likelihood (FISHER):

$$\mu = \frac{\Sigma M_i}{\Sigma N_i t_i} \qquad \text{(one-hit hypothesis)}$$

$$\mu^2 = \frac{\Sigma M_i}{\Sigma N_i t_i^2} \qquad \text{(two-hit hypothesis)}$$

$h_i = M_i/N_i$ = mutants per total number of phages.

The experimentally obtained h-values and the theoretical p-value can be tested for their level of significance with the Chi^2 test.

Literature

SMITH, K.C., HANAWALT, P.C.: Molecular Photobiology, p. 230. New York: Academic Press 1969.

TIMOFEEFF-RESSOVSKY, N.W., ZIMMER, K.G.: Das Trefferprinzip in der Biologie, 317 p. Leipzig: Hirzel Publ. 1947.

14. Reversible Photochemical Alteration of Cytidylic Acid

When nucleic acids or their constituents are irradiated with
UV light, many different photoproducts, especially reaction
products of the pyrimidines, are formed. The best-known photochemical processes are:

- Dimerization, e.g. $T + T \xrightarrow{h\nu} \widehat{TT}$

- Deamination, e.g. $C \xrightarrow{h\nu} U$

- Hydrate formation (see this Expt.)

- Cross linking, e.g., of nucleic acids with protein.

Radiochemical and radiobiological tests have shown that a large
portion of the inactivating and mutagenic reactions induced by
UV irradiation of biological specimens is probably based on
dimerization. However, this statement does not explain the
biological UV effect at the molecular level, because there are
more than 10 different ways for pyrimidines to form dimers but
not every type of dimer must have biological consequences.

Effective formation of dimers in pyrimidine solutions is only
possible when the samples are irradiated in a frozen state.
Because the UV-induced formation of hydrates of cytidylic acid
occurs at room temperature, this reaction will be used in the
present experiment to demonstrate certain characteristics of
photochemical processes.

If cytidylic acid (= Cp = cytidine monophosphate) is irradiated
with UV light, its characteristic absorption maximum at 270 nm
disappears and a new one is found at 240 nm. The photoproduct
which has been formed by irradiation (Cp*) is unstable and most
of it reverts to cytidylic acid at room temperature. A small
portion (<10%) of Cp* deaminates spontaneously to the corresponding derivative of uridylic acid (Up*). The photoproduct Cp* is
probably formed by hydration at the 5, 6 double bond:

Ribose phosphate   Ribose phosphate

Photochemical processes are in general characterized by three
parameters:

a) The effective cross section, gives the probability with
which a certain "event" takes place at a given photon flux, D,

e.g., the change of Cp to Cp*, which can be recognized by the decrease of the extinction (E) at 270 nm.

$$\sigma = \frac{(\log) E_0 - \log E) \times 2.30}{D} \left[\frac{cm^2}{\mu E} \right]. \tag{1}$$

The photon flux, D ("irradiation dose"), is measured in micro-Einsteins per cm^2 and must not be mistaken for the photon-flux rate, which is defined as photon flux per time unit. 1 micro-Einstein (μE) corresponds to 1 μmole of photons or 6.02×10^{17} photons. Eq. (1) is derived from the exponential equation

$$E/E_0 = e^{-\sigma D}. \tag{1a}$$

(e = natural base of logarithms)

b) The absorption cross section, s gives the probability with which a photon is absorbed at a given photon flux.

$$s = 2.30 \times 10^{-3} \times \varepsilon_\lambda \left[\frac{cm^2}{\mu mole} \right]. \tag{2}$$

(For details see JOHNS, 1969).

Here ε_λ is the molar extinction coefficient of the unirradiated chemical substance for a given wave length, λ.

$$\varepsilon_\lambda = \frac{E_\lambda}{c \times 1} \left[\frac{cm^2}{mmole} \right]. \tag{3}$$

   c: Concentration of the chemical substance in mole/l or m mole/cc
   1: light path in cm
   E: log D_0/D

c) The quantum yield, Φ, is given by the ratio of the effective cross-section σ and the absorption cross section s, i.e., it sets the photochemically altered molecules in relation to the number of photons absorbed. Alternatively, it can be said that Φ is the ratio of photochemically effective photons to the number of all photons absorbed.

$$\Phi = \frac{\sigma}{s} \left[\frac{\mu mole}{\mu E} \right]. \tag{4}$$

The quantum yield of many photochemical reactions in pyrimidine bases and amino acids lies between 10^{-2} and 10^{-4} μmole/μE.

Plan. The UV-induced change from cytidylic acid (Cp) into the corresponding hydrate (Cp*) will be photometrically followed as a function of the irradiation time. Assuming that the UV-lamp used emits monochromatic light (254 nm), the cross section σ and the quantum yield Φ for the reaction Cp $\longrightarrow$ Cp* will be calculated. The spontaneous back reaction Cp* $\longrightarrow$ Cp, will be photometrically followed and the fraction of the photoproduct which can revert will be estimated.

Material. 150 ml of a 6 × 10⁻⁵ M solution of cytidine monophosphate (MW = 323.2) in 5 mM SORENSEN-phosphate buffer, pH 8.0.
9 covered glass petri dishes (diameter about 8.8 cm) with a plain
bottom. 1 magnetic stirrer with a very thin magnetic rod. 1 ozone-
free Hg low pressure lamp, e.g., OSRAM HNS 12, without metal housing, horizontally fixed on a stand. 2 pairs of plastic gloves.
1 sheet each of graph paper and semi-logarithmic paper. For several groups together: 1 UV spectrophotometer with 3 quartz cuvettes
of 1 cm light path.

Procedure

As UV light causes skin burns, avoid exposing skin and use
protective goggles and plastic gloves.

1. Irradiation (data sheet I). Switch on the UV lamp 15 min
before experiment begins and keep it on during the whole experiment. Irradiate 14 ml of a Cp solution in a petri dish at
a distance of 3 cm from the UV lamp for 1 min and stir magnetically. The exact irradiation time can be controlled by opening
and closing the petri dish with a piece of cardboard. Immediately
afterwards measure the extinction at 270, 254, and 240 nm against
P-buffer. Carry out all other irradiations in the same way with
fresh 14 ml of Cp solution (non-fractionated irradiation).

2. Absorption spectrum (data sheet II). Do not turn off the stop
watch after 20 min irradiation! Determine the extinction at 270,
254 and 240 nm and then measure the absorption spectrum between
220 and 300 nm at intervals of 5 nm. Along with the sample irradiated for 20 min, also determine the spectrum of the non-irradiated Cp solution.

3. Follow photometrically the spontaneous reversion of Cp* ⟶ Cp
(data sheet III) immediately after the sample was irradiated for
20 min. Measurements at 270 and 240 nm are sufficient. Measure
first in intervals of 5 or 10 min and when the reaction has
slowed down take the readings at 30 min intervals, the last
reading after 6 hrs.

Evaluation

Plot the results of data sheets I and III semi-logarithmically
and those of data sheet II linearly. All of the following calculations give only approximative values.

1. Calculate the extinction coefficient ε according to Eq. (3)
for Cp and Cp* (≈20 min UV-irradiated Cp). The Cp solution is
6 × 10⁻⁵ M.

	ε_{240}	ε_{254}	ε_{270}
Cp			
Cp*			

2. Calculate the <u>effective cross section</u> σ from the linear part of the reaction Cp⟶Cp* measured at 270 nm according to Eq. (1). The photon-flux rate is about 1 μE cm^{-2} min^{-1} under the given test conditions. (A simple approximation method for the determination of the photon-flux rate is given in the section "Experimental Results").

$$\sigma = \frac{\rule{3cm}{0.4pt}}{} = \frac{}{\rule{2cm}{0.4pt}} \quad \left[\frac{cm^2}{\mu E}\right]$$

3. Calculate the <u>quantum yield</u> Φ for the reaction Cp⟶Cp* according to Eq. (2). Use the extinction coefficient ε_{254} obtained above in section 1, as the UV lamp used emits mainly radiation of the Hg-resonance line 253.7 nm.

$$\Phi = \frac{\rule{3cm}{0.4pt}}{} = \frac{}{\rule{2cm}{0.4pt}} \quad \left[\frac{\mu mole}{\mu E}\right]$$

4. <u>Reversion</u> of Cp*⟶Cp. What percentage of the photoproduct spontaneously reverts within 6 hrs, assuming that all Cp has been changed into Cp* after 20 min UV radiation?

Literature

BECKER, H., LeBLANC, J.J., JOHNS, H.E.: The UV Photochemistry of Cytidylic Acid. Photochem. Photobiol. <u>6</u>, 733-743 (1967).

JAGGER, J.: Introduction to Research in Ultraviolet Photobiology. Englewood Cliffs, N.J.: Prentice-Hall 1967.

JOHNS, H.E.: Photochemical Reactions in Nucleic Acids. In: Methods in Enzymology <u>16</u>, 253-316 (1969).

WANG, S.Y. (ed.): Photochemistry and Photobiology of Nucleic Acids, Vol. 2. New York: Academic Press 1976.

<u>Time requirement:</u> 1st day 7.5 hrs.

Data sheet I: Cp⟶Cp*

Time	Extinction at λ (nm)		
min	270	254	240
0 1			
2 3			
4 6			
8 10			
15 20			

Data sheet II: Absorption spectrum of Cp and Cp*

Wave length	Extinction Cp	Cp*	Wave length	Extinction Cp	Cp*
220 225			265 270		
230 235			275 280		
240 245			285 290		
250 255			295 300		
260			305		

Data sheet III: Cp*⟶Cp

Time	Extinction at λ (nm)		Time	Extinction at λ (nm)	
min	270	240	min	270	240

15. Chemical Induction of Forward and Back Mutations in Bacteria

In the forties it was discovered that certain chemicals induce
mutations. Mutagenic chemicals are used:

- To induce mutations. Because of their functional or structural
defects, mutant organisms are of wide-spread use, e.g. for the
study of biosynthetic pathways (compare Expt. 22) or of morpho-
poietic processes (compare Expt. 25). Chemical mutagens are
easier to handle and less expensive than many physical ones.

- To investigate the mechanism of mutation. It was found that
the reaction of many chemical mutagens is very specific, making
it possible to predict the type of mutation. For example, acridine orange can only cause "frame shift" mutations in the phage
DNA in T4-infected bacteria. The base analogue, 5-bromodeoxyuri-
dine, on the other hand, only induces base-pair exchanges (substitution of pyrimidines).

The following experiment has been divided into two parts:
First, wild-type bacteria will be treated with nitrosoguanidine
by a routine laboratory method. Nitrosoguanidine (NG) is the
strongest mutagen among those presently available: N-methyl-N'-
nitro-N-nitrosoguanidine

(NG)

$$O_2N \longrightarrow N = C \longrightarrow N \Big\langle \begin{matrix} N = O \\ CH_3 \end{matrix}$$
$$| \\ NH_2$$

In bacteria it induces transitions and transversions, but no
frame shift mutations. Although NG can alkylate, the question of
whether its mutagenicity is due to this reaction is still under
investigation. Different ways for the selection of mutants can
be followed.

In the second part, a "filter disc method" will be applied, which
(with various modifications) allows a rapid screening of chemicals for their mutagenicity. Furthermore, this method is frequently used to test whether certain chemicals can revert bacterial and phage mutants to "wild-type" (see below). From the
results of such tests conclusions may be drawn as to the original forward-mutation and to the base specificity of a yet unknown chemical mutagen. For example, mutations of the frame-
shift type can only be reverted by new frame-shift mutations and
not by base-pair exchanges. In this context, it should be kept
in mind that many revertants, although phenotypically wild-type,
do not contain the base-sequence of the wild-type. These revertants can arise by suppressor mutations, which occur at a site
different from the site of the forward mutation and which may
cancel the effect of the forward mutation to various extents.

Plan. 1. Growing cells of *Serratia marcescens*, wild type, will be treated with nitrosoguanidine under defined conditions, in order to induce <u>forward mutations</u> to auxotrophy and to deficiency of pigment synthesis. The frequency of mutants will be determined immediately after mutagenesis and after a post-incubation of several hours. Thus the segregation of mutated and non-mutated alleles can be observed.

2. With the "filter disc method" the spontaneous and the chemically induced <u>back mutations</u> of an auxotrophic bacterial mutant will be investigated.

Material. 1. Forward mutation. 10 ml of a log culture in NB ($30^{\circ}C$, aerated) of *Serratia marcescens* W225 (wild type); titer 3×10^8/ml. 15 NB agar and 15 M9 agar plates. 1 block of wood and 15 sterile velvet cloths for replica plating (for more information, see Expt. 11). 1 ml of N-methyl-N'-nitro-N-nitrosoguanidine, 100 µg/ml, dissolved in P-buffer, of pH 6.0. 10 ml of the same P-buffer, with no supplements. 25 ml NB. 2 10-ml centrifuge tubes. For several groups together: 1 SORVALL centrifuge RC2-B with an SS34 rotor, 1 vortex mixer.

2. <u>Back mutation</u>. 1 ml of a stat culture, in NB ($30^{\circ}C$, aerated), of *Serratia marcescens* W366 (<u>thi</u>), adjusted with P-buffer to a cell density of approx. 6×10^8/ml. 3 M9s agar plates. 3 soft agar tubes without NB. All of the following solutions are prepared with distilled water. 2 ml of 0.2% 5-bromodeoxyuridine (BUdR); 2 ml of 1% deoxythymidine (dThy); 2 ml of 1% deoxyadenosine (dAde); 2 ml 2% nitrosoguanidine); 1 ml of 10% hydroxylamine. 1 pair of pointed forceps. 1 petri dish with sterile filter discs (for details, see material of Expt. 22). 1 beaker with about 200 ml tap water. 2 paper towels.

Procedure

Caution! Avoid skin contact with mutagens. Never pipette mutagen solutions with the mouth - use a propipette instead.

1. Forward mutation. 1st day. Following data sheet I, prepare tubes for dilution with P-buffer and number 15 NB agar plates. Centrifuge 10 ml of a log culture of *S. marcescens* W225 at 4,500 rpm ($2,500 \times g$) for 10 min and resuspend the pellet in 5 ml P-buffer at pH 6.0 with a vortex mixer. Microscopically determine how many of the cells exist in pairs or have formed clumps (magn. 10×40). Immediately mix 0.5 ml of the bacterial suspension with 0.5 ml NG-solution (t = 0) and incubate at $30^{\circ}C$. Following data sheet I, take 0.1 ml samples at t = 0 and 30 min, dilute in P-buffer and spread on NB agar plates with a glass rod. Immediately after removal of the 30 min sample, add 9 ml ice-cold NB to the remaining NG-cell suspension, in order to lower the NG-concentration, and centrifuge at 4,500 rpm ($2,500 \times g$) for 5 min. Discard the supernatant, resuspend the sediment with a vortex mixer in 10 ml of NB and aerate at $30^{\circ}C$ until a cell titer of approx. 4×10^7/ml is reached. Again take a sample, dilute and spread on NB agar plates (data sheet I). Incubate all plates at $30^{\circ}C$ for at least 20 hrs.

2nd day: Count the colonies on the NB agar plates Nos. 1-15 and distinguish carefully between the four different classes indicated on the data sheet I. Then transfer the colonies with sterile velvet cloths to M9 agar plates (having identical plate numbers) and incubate at 30°C for about 20 hrs.

2. Back mutation (filter-disc method). 1st day. Number three M9s-agar plates on the bottom sides and divide into 4 fields (A, B, C, D) of identical size. On each plate pour 0.1 ml of a stat culture of S. marcescens W366 (thi) suspended in 3 ml each of soft agar without NB. Mix equal volumes (0.3 ml each) of the solutions of mutagens or bases and distilled water in small test tubes as indicated in data sheet III. Then dip a sterile filter disc into the BUdR solution and put it on plate No. 1, position A. Avoid excess solution on the filter discs. Finally, wash forceps in tap water and dry with a paper towel. Repeat with solutions 7-9 in the same manner, according to data sheet III. Incubate all three agar plates at 30°C for 2-3 days.

To save time, the filter-disc method can be carried out in those 2-3 hrs when the NG-treated wild type bacteria are being incubated for segregation (see Part 1: Forward mutation).

Evaluation

1. Forward mutation. 3rd day (see data sheet II)

- Survival of S. marcescens W225 after 30 min NG-treatment without post-incubation:

$$\frac{\text{(Colonies on NB agar plates 6 to 10)} \times 100}{\text{(Colonies on NB agar plates 1 to 5}} = \underline{\qquad} = \underline{\qquad} \%$$

- Frequency of mutants as percentages with standard deviation (see nomogram on page 12). Pool the results of several student groups if you are sure that the criteria applied by different groups to classify a colony are always about the same.

- Question: How can one interpret the differences in the mutant frequencies between cells immediately plated after the NG-treatment and those which were post-incubated in nutrient broth before plating?

2. Back mutation (filter disc method). 3rd day

- Estimate the number of back mutant colonies surrounding each filter disc in a radius of about 1.5 cm, and record on data sheet III. Which of the tested substances is mutagenic? Why is the mutagenicity of a substance abolished in presence of a certain DNA base?

- Which of the tested substances inhibits bacterial growth at high concentration? Does a correlation exist between growth inhibition and mutagenicity?

Data sheet I

Time t (min) of mutagen treatment	Dilution	Plate No.	Colonies on NB agar					Auxotrophs[a]
			Red	White or pink	Color sectoring	Others	Sum	
0	5×10^{-6}	1 2 3 4 5						
30 min without post-incubation	5×10^{-6}	6 7 8 9 10						
30 min with post-incubation	2×10^{-5}	11 12 13 14 15						

[a] Colonies transferred by the velvet stamp to minimal medium and which did not continue to grow.

Data sheet II

t (min)	Auxotrophic mutants	Pigmentation mutants	
		Color-sectorized colonies	Uniformly colored ("pure") colonies[a]
O			
30 (no incub.)			
30 (incub.)			

[a] "Pure" color-mutant clones are the uniformly white or pink colored colonies here. Their definite classification as mutants would require subculturing in order to see which of the colony colors that has deviated from the wild type is actually inheritable.

Data sheet III

Plate No.	A	B	C	D
1	BUdR + d.w.	BUdR + dThy	BUdR + dAde	dist. water
2	NG + d.w.	NG + dThy	NG + dAde	dist. water
3	HA + d.w.	*	*	dist. water

* Solutions or crystals of other mutagens, e.g., 2-aminopurine, mitomycin C or ICR-191.

Literature

FISHBEIN, W., FLAMM, G., FALK, H.L.: Chemical Mutagens. Environmental Effects on Biological Systems, p. 306. New York:
  Academic Press 1970.

HOLLAENDER, A. (ed.): Chemical Mutagens. Principles and Methods
  for Their Detection. Vol. I and II. London: Plenum Press 1971.

KAPLAN, R.W.: Probleme der Prüfung von Pharmaka, Zusatzstoffen
  u.a. Chemikalien auf ihre mutationsauslösende Wirkung.
  Naturwissenschaften 49, 457-462 (1962).

Time requirement: 1st day 4.5 hrs, 2nd day 1 hr.

Problems

1. Many mutagens are "base specific", i.e., they change or substitute only certain purines and/or pyrimidines. The base specificity of the following mutagens should be characterized in
the table by (+) and (-) signs.

Base	BUdR	HNO$_2$	NH$_2$OH	2AP	UV
A					
T					
G					
C					
Molecular mechanism of action					

BUdR   : 5-bromodeoxyuridine   2AP : 2-aminopurine

HNO$_2$: Nitrous acid   UV : Ultraviolet light

NH$_2$OH : Hydroxylamine

2. The following 6 protein segments are homologous. The mutant
proteins originated from independent single point mutations in
the wild type. On the basis of the genetic code (see p. 233)
note the different mRNA base sequences which result from the
corresponding amino acid exchanges. Then answer the following
questions:

a) How many different mutations could explain each single
amino acid exchange?
b) Of what type would these mutations be (transition, trans-
version, insertion, deletion)?

Protein from	Amino acid sequence			
Wild type	--- phe	ser	pro	trp ---
Mutant 1	--- leu	ser	pro	trp ---
Mutant 2	--- cys	ser	pro	trp ---
Mutant 3	--- phe	ser	pro	- ---
Mutant 4	--- phe	ser	pro	gly ---
Mutant 5	--- phe	ser	ala	leu ---

The exchanged amino acids are underlined for emphasis.

3. Assuming the spontaneous mutation rate for a$^+$ $\longrightarrow$ a were
α = 2 × 10^{-6} mutations per cell per generation, at what cell
number (N) would a growing culture then reach its "critical
population size" N_K, (= one mutation per culture, on the
average)?

$$N_K = \ln 2/\alpha$$

D. Transfer and Recombination of Genetic Material

"Recombination" means the formation of new genotypes by exchange
of genetic material from two parents differing in two or more
inheritable properties. In the genetics of higher organisms we
differentiate between

a) Recombination of "linked" genes, which lie on homologous
chromosomes and can only be recombined by a crossover, and

b) Recombination of "non-linked" genes, which lie on non-homologous chromosomes and therefore can be freely combined without
a crossover.

Both types of recombination are found in *E. coli* bacteria. Recombination of linked genes predominates because all of the
genetic information needed for the reproduction of *E. coli* is
arranged on a single, circular chromosome of high molecular
weight (approx. 3×10^9). Recombination of non-linked genes is
sometimes possible, because certain bacteria contain one or more
plasmids in addition to their chromosomes. Plasmids are circular
DNA molecules of relatively low molecular weight. A plasmid can
be essential for a bacterium depending on its genetic information and the environmental conditions.

The following parasexual (i.e. ameiotic) processes make recombination in bacteria possible:

1. Conjugation (Expt. 16). When suitable cells are in contact,
a copy of the chromosome of a "donor" cell is transferred,
partially or entirely, to a "recipient" cell.

2. Transduction (Expt. 18). Small fragments of bacterial chromosomes are transferred by phages from one cell to another. No
direct cell contact is necessary for this process.

3. Transformation. Cells in a "competent" state are able to
take up DNA which has been isolated from other cells. Cells
of certain bacterial strains can be made "competent", which
means that they are made able to take up free DNA. For successful transformation the DNA must consist of chromosomal fragments
isolated from the same or a related bacterial strain.

4. Plasmid transfer and sexduction (Expt. 19). By contact of
suitable cells, a copy of a donor cell plasmid migrates to a
recipient cell. In sexduction, the plasmids contain, in addition to their own genes, genes of the bacterial chromosome.

In order to gain stable recombinants through parasexual processes 1-3, crossovers between the newly introduced DNA (exogenote) and the resident DNA must occur. A non-integrated exogenote cannot be replicated autonomously and would be passed on to only one of the two daughter cells after each cell division. In plasmid transfer (4) no crossovers are necessary in order to gain a stable recombinant, since the plasmid may replicate autonomously.

There are several loci on the *E. coli* chromosome at which mutations can occur which render cells recombination-deficient. Many of these so-called <u>rec</u> mutants are more sensitive to ultraviolet light than the wild type. Thus it can be assumed that there must be bacterial enzymes which participate in recombination as well as in the repair of radiation damage in DNA. This assumption is supported by the fact that weak UV irradiation of the cells increases the chances of recombination. To date, the precise mechanism of genetic recombination remains still unclear. It is assumed that specific endo- and exonucleases, a ligase and perhaps other enzymes are involved in the recombination process.

Phage DNA can also recombine. If bacteria are mixedly infected with homologous phages which differ in certain genetic markers, the phage progeny contain recombinants as well as the parental genotypes (Expt. 17).

Analogous to the situation in bacteria, certain phages (e.g. λ) also have genes whose products promote phage recombination. If one of these genes mutates, the phages do not lose the ability to recombine as long as bacterial "recombinases" are available.

The integration of the DNA of temperate phages into the host chromosome is called "lysogenization". Lysogenization, as well as the excision of a prophage during induction (Expt. 20), are processes involving crossovers.

Literature

CLARK, A.J.: Recombination Deficient Mutants of *E. coli* and Other Bacteria. Ann. Rev. Genet. 7, 67-86 (1973).

CURTISS, R. III: Bacterial Conjugation. Ann. Rev. Microbiol. 23, 69-136 (1969).

DAVERN, C.J.: Molecular Aspects of Genetic Recombination. In: Progress in Nucleic Acid Research and Molecular Biology 11, 229-258 (1971).

GROSSMAN, L., MOLDAVE, K. (eds.): Methods in Enzymology, 21D, Section V, Enzymes Involved in Recombination and Replication, pp. 289-338, and Section VI, Gene Localization Techniques, pp. 341-480 (1971).

HOTCHKISS, R.D.: Models of Genetic Recombination. Ann. Rev. Microbiol. 28, 445-468 (1974).

RADDING, C.M.: Molecular Mechanisms in Recombination. Ann. Rev. Genet. 7, 87-111 (1973).

WHITEHOUSE, H.L.K.: The Mechanism of Genetic Recombination. Biological Reviews 45, 265-315 (1970).

16. Conjugation of *E. coli* Bacteria

There are two primary characteristics of the conjugation of *E. coli* bacteria. First, cell-to-cell contact of the conjugating individuals is required. Second, the transfer of genetic material (DNA) is only in one direction, namely from a "donor" to a "recipient". Donors are those bacteria which contain a fertility factor (F factor). The F factor is a DNA molecule with a molecular weight of approx. 6×10^7 daltons. In *E. coli* three fertility types are known.

1. F⁻ cells ("females"). They have no F factor and can only function as recipients.

2. F⁺ cells. They contain a few circular, autonomously replicating copies of F factors in the cytoplasm. If F⁺ and F⁻ cells come into contact, the F factor will be transferred with a high frequency (see Expt. 19). In every F⁺ population a few Hfr cells can be found, which have formed spontaneously from F⁺ cells (see below).

3. Hfr cells ("males"). A single F factor is linearly integrated into the circular chromosome to form an Hfr cell and is replicated with the chromosome. Hfr cells transfer their bacterial chromosomes with a high frequency to recipient cells. Hfr stands for high frequency of recombination.

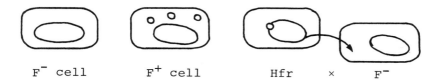

F⁻ cell   F⁺ cell   Hfr   ×   F⁻

Thread-like filaments, the sex or F pili, which are not found on F⁻ bacteria, have been observed on Hfr and F⁺ cells and seem to be necessary for cell contact with a mating partner. During the conjugation of Hfr × F⁻, a linear duplicate of the Hfr chromosome is probably "pushed" through such a pilus into the F⁻ cell. It takes about 2 hrs for the complete Hfr chromosome to be transferred to the F⁻ cell. In general, the transfer is interrupted before it is completed because cell contact of the partners is very sensitive to shearing forces. The recipient cell then gets only a fragment of the donor chromosome. This fragment can be "recombined" into the recipient's chromosome, either completely or partially. The homologous DNA segment on the recipient's chromosome is eliminated; for this reason no reciprocal recombinants occur simultaneously. The process of recombination, probably an interplay of several nucleases, replicases and ligases, is under the control of several rec genes.

Different Hfr strains transfer their genes into a recipient cell in a different sequence; for example, A-B-C-D or D-C-B-A or even C-D-A-B. This phenomenon is based on the fact that the F factor can be inserted at different sites on the chromosome. The site of insertion determines where the circular chromosome will open and which gene will be transferred first, i.e., the origin of transfer (o). The F factor itself, or part of it, always migrates at the end of the chromosome into the recipient, and will only be transferred in conjugation events which are not interrupted prematurely.

With the help of conjugation (and also transduction) more than 400 genes were localized on the *E. coli* chromosome. Along with *E. coli*, conjugations occur also in other bacterial genera, e.g., in Salmonella, Pseudomonas and Rhizobium.

Plan. Two strains of *E. coli* will be conjugated. One is a prototrophic, streptomycin-sensitive Hfr strain and the other a multiply auxotrophic, streptomycin-resistant F⁻ strain. At varying times after the beginning of the conjugation samples will be taken and the conjugation pairs which have formed will be separated by vigorous shaking. The samples will be plated onto different selective minimal media, whose streptomycin content allows no growth of the prototrophic Hfr cells. The time interval between the entrance of single selected Hfr markers (arg^+, pro^+, thr^+, leu^+) into the F⁻ cells will be determined from the increase in different recombinant types as a function of the duration of conjugation. These time intervals and the sequence of the markers are a relative measure of their degree of linkage. The markers thr^+ and leu^+ will be used as one marker since they are very closely linked.

Material. 4 ml of a log culture, in TBY broth (aerated at 37°C), of *E. coli* AB1157 F⁻ (thr leu arg pro his thi str^r), and others (see Appendix B)). 2 ml of a log culture, in TBY broth (not aerated at 37°C), of *E. coli* GY767 Hfr (str^s). The cell titer of both cultures should be approx. 3×10^8/ml.

	Additions
8 M9 agar plates (M9-A)	Str, Thi, His, Arg, Thr, Leu
12 M9 agar plates (M9-B)	Str, Thi, His, Arg
8 M9 agar plates (M9-C)	Str, Thi, His

Concentrations – amino acids: 50 µg/ml; streptomycin sulfate: 40 µg/ml; and thiamine-HCl: 0.5 µg/ml. 250-ml Erlenmeyer flask with a lead ring (to stabilize it in the water bath). 37°C water bath. 10 ml of TBY broth. 1 vortex mixter. 1 sheet of linear graph paper.

Procedure

1. Preparation. Number small test tubes 1-14 and fill each with 0.9 ml P-buffer. Label all agar plates, following the data sheet.

2. <u>Pair formation</u>. Warm 2.7 ml of a culture of AB1157 in a
250-ml Erlenmeyer flask (weighted with lead ring) in the water
bath at 37°C.

At t = 0 min, add 0.3 ml of a culture of GY767. Mix for a
short time, then carefully incubate. Ideally
there should be no vibration.

At t = 5 min, carefully add 7 ml of TBY broth, prewarmed to
37°C. This dilution lessens the chance of the
formation of still further conjugation pairs
after 5 min.

3. <u>Pair separation</u>

At t = 6 min, add a 0.1 ml sample from the conjugation mixture
to test tube 1 (containing 0.9 ml P-buffer) and
shake vigorously with mixer for at least 30 sec,
dilute 10^{-1} in P-buffer and spread, following the
data sheet, 0.1 ml samples of this suspension onto
different agar plates. Incubate plates at 37°C for
48 hrs.

At t = 9 min, 12 min, etc. (see data sheet) proceed as at t = 6 min.

Selection will be made according to the donor gene markers:

pro^+ on M9 A agar

pro^+ thr^+ leu^+ on M9 B agar

pro^+ thr^+ leu^+ arg^+ on M9 C agar

<u>Evaluation (3rd day)</u>

1. Count the colonies on the M9 plates and record on the data
sheet. Plot the number of colonies as a function of time from
beginning of conjugation (linear graph paper). Extrapolate
the linear part of the curve back to the point of intersection
with the time axis. These intersections show the earliest time
at which each of the Hfr markers investigated have entered the
F^- cells.

O	10	20	30	40	50 min

.... Markers

2. Calculate the ratio of the amount of Hfr to F^- cells in the
conjugation mixture.

Hfr/F^- =

Why is this ratio smaller than 1?

3. Calculate the frequency of the recombinants with respect to
the Hfr cells in the conjugation mixture. For the numerator,
use the highest colony number which was obtained on M9 plates.

The denominator is obtained from the titer of the GY767 culture taking into account the dilution steps which preceded the spreading.

Literature

CURTISS, R., III: Bacterial Conjugation. Ann. Rev. Microbiol. 23, 69-136 (1969).

HAYES, W.: The Genetics of Bacteria and their Viruses, 925 p. (esp. pp. 650-699). Oxford: Blackwell Scient. Publ. 1968.

JACOB, F., WOLLMAN, E.L.: Sexuality and the Genetics of Bacteria. New York: Academic Press 1961.

LEVINTHAL, M.: Bacterial Genetics Excluding *E. coli*. Ann. Rev. Microbiol. 28, 219-230 (1974).

TAYLOR, A.L., TROTTER, C.D.: Revised Linkage Map of *E. coli*. Bacteriol. Rev. 31, 332-353 (1967).

TAYLOR, A.L., TROTTER, C.D.: Linkage Map of *E. coli* Strain K12. Bacteriol. Rev. 36, 504-524 (1972).

WOOD, T.H.: Effects of Temperature, Agitation, and Donor Strain on Chromosome Transfer in *E. coli*. J. Bacteriol. 96, 2077-2084 (1968).

Time requirement: 1st day 2 hrs, 3rd day 2.5 hrs.

Data sheet

Time t of sample taken (min)	Plate No.	Colonies per plate on		
		M9 A	M9 B	M9 C
6	1			
9	2			/
12	3			/
15	4			/
18	5			/
21	6			/
24	7			/
30	8			
36	9	/		
42	10	/		
48	11	/		
54	12	/		
60	13	/	/	
70	14	/	/	

/ means that no plates were streaked.

17. Two- and Three-Factor Crosses with Phage Kappa

If bacteria are infected simultaneously with two phages which
differ in at least two genetic "markers", e.g., (x +) and (+ y),
among the phage progeny, wild types (++) and double mutants (xy)
will be found as well as both parental genotypes. Thus, phages,
like higher organisms, can exchange genetic material. In phages
this exchange results from enzymatic "breakage and reunion" of
the parental DNA molecules.

The probability for crossovers between two linked genetic markers,
increases with the distance between them as shown, for example,
by deletion mapping. The frequency of recombinants reflects the
frequency of crossovers and is a relative measure for the distance between genes on a chromosome. If a number of genetic
markers which are distributed over the entire genome are crossed
in pairs, the distances between these markers are approximately
additive. The genetic markers can thus be arranged on linear or
circular "genetic maps".

During phage replication in a mixedly infected host cell all the
phage chromosomes can participate in one or several independent
recombination events. This means that formation of phage recombinants is an event of population genetics, in contrast to the
meiotic processes in higher organisms. The appearance of recombinants depends on the number of genomes and the time during
which they remain in the DNA pool. Evidence supporting this con-
cept is as follows:

(1) If the lysis of mixedly-infected bacteria is delayed, the
frequency of recombinants increases because, first, the DNA pool
per cell is enlarged, and second, there is "more time" for recombination. (2) If bacteria are infected simultaneously with 3 instead of the usual two genetically different phages, recombinants
are found in which markers of all three parents are combined.
Such recombinants probably arise from two sequential crossovers.

If a population of mixedly-infected bacteria is lysed both of
the products of the reciprocal recombination are found among
the total phage progeny at about the same frequency. On the
other hand, the phage progeny of single bursts frequently contain the reciprocal recombinants in unequal amounts. One explanation for this is that during the intracellular maturation of
the phages the DNA molecules are statistically withdrawn from
the pool of continuously replicating and recombining phage genomes. Furthermore, many genomes never mature to infectious

particles and are thus lost for genetic analysis. Another hypothesis states that all recombinants, when formed, go through a heteroduplex-heterozygote stage; for example:

```
    x                    y
_____  - - - - - - - - - -
_____  - - - -
    x                    +
```

The replication of these intermediate DNA molecules could result in an unequal frequency of reciprocal recombinants. Some of these intermediate products also appear as mature phages, the so-called HETs. In phage T4 and others, additional "redundancy heterozygotes" are found (for details see MOSIG). Another phenomenon frequently found in phage crosses is <u>Negative Interference</u>. This means that a single crossover increases the probability of further crossovers in the immediate vicinity. <u>Positive Interference</u> is the opposite process. A measurement of interference is the <u>Coincidence Factor</u>, C.

$$C = \frac{F_{double}}{F_I \times F_{II}}$$

C > 1 negative interference

C = 1 no interference

C < 1 positive interference

F_{double} : The experimentally observed frequency of a recombinant type, which can only be formed by an even number of crossovers.

$F_I \times F_{II}$: The theoretically expected frequency of the same recombinant type, if the necessary crossovers would occur independently of each other and at random.

<u>Plan</u>. The order of three linked genetic markers x, y and z (nonsense triplets) of phage Kappa will be determined by two-factor crosses x × y, y × z and x × z. A suppressor strain (su^+) of *Serratia marcescens* is infected with 2 different nonsense (<u>sus</u>) mutants of the phage, and incubated until lysis of the bacteria occurs. If the phage progeny are titered with su^+ bacteria, the phages of <u>all</u> 4 genotypes form plaques (e.g., x+, +y, ++, xy). However, only the (++) recombinants form plaques if titered on non-suppressing (<u>su</u>) bacteria. The gene order (preliminary maps) which has been found with these experiments is to be verified by the 3-factor cross xz × y.

<u>Material</u>. 5 ml each of stat cultures, in NB (30°C, aerated) of *Serratia marcescens* <u>su</u> W225 and <u>su</u>$^+$ W319 as indicator; cell titer approximately 1×10^9/ml. 3 ml of a log culture of *S. marcescens* <u>su</u>$^+$ W319 with a cell titer of 3×10^8/ml. 0.5 ml (accurately!) of a 1:1 mixture of the parental phages x + y, y + z, x + z and xz + y. The plaque titers of these mixtures of <u>sus</u> mutants of the phage Kappa is 3.6×10^9/ml on <u>su</u>$^+$ W319 indicator and less than 3.6×10^4/ml on <u>su</u> W225 indicator, i.e., the portion of spontaneous back mutants is $<10^{-5}$. 60 ml of NB. 16 plates with NB agar. 16 tubes with NB soft agar. For several groups together: 30°C water bath.

Procedure (1st day)

Preparation. Following data sheet I, label NB agar plates and prepare dilution rows with P-buffer. In addition, pipette 9.9 ml and 1.8 ml NB into 4 test tubes each and label xy, yz, xz and xzy.

Phage cross. Pipette 0.5 ml of the log cells of $\underline{su}^+$ W319 into the above phage mixtures which have been labeled xy, yz, xz and xzy. Incubate these mixtures for adsorption of the phages in the 30°C water bath for 5 min and then dilute in NB to 10^{-3} (mix 0.1 ml phage suspension + 9.9 ml NB and from this mixture take 0.2 ml + 1.8 ml NB). Keep the four 2-ml test tubes in the 30°C water bath a further 60 min so that the phages will replicate and the host cells will lyse. Add approximately 5 drops of chloroform to each mating mixture, shake for a moment and leave at room temperature for 10 min. Those bacteria which have not yet lysed will then rupture without damage to the phages. Dilute the "phage lysates" obtained, following data sheet I and plate with $\underline{su}^+$ W319 and $\underline{su}$ W225 as indicators (0.1 ml sample + 0.2 ml indicator + 3 ml soft agar per plate). Incubate agar plates at 30°C overnight.

Data sheet I

Phage progeny of	Dilutions	Indicator $\underline{su}$ W225 Plate No.	Plaques	Indicator $\underline{su}^+$ W319 Plate No.	Plaques
x × y	1×10^{-2}	1		–	
	1×10^{-2}	2		–	
	3×10^{-4}	–		3	
	3×10^{-4}	–		4	
y × z	1×10^{-2}	5		–	
	1×10^{-2}	6		–	
	3×10^{-4}	–		7	
	3×10^{-4}	–		8	
x × z	1×10^{-2}	9		–	
	1×10^{-2}	10		–	
	3×10^{-4}	–		11	
	3×10^{-4}	–		12	
xz × y	1×10^{-1}	13		–	
	1×10^{-1}	14		–	
	3×10^{-4}	–		15	
	3×10^{-4}	–		16	

Evaluation (2nd day)

Count the plaques on agar plates Nos. 1-16. Calculate separately for all four phage lysates and list on data sheet II:

1. The plaque titer on bacterial hosts su W225 and su⁺ W319.
Remember that the mixtures of phage and bacteria for the 4 phage
crosses were diluted twice, first 10^{-3} in NB and then once more
in buffer 10^{-1}, 10^{-2} or 3×10^{-4}, respectively.

2. The frequencies of recombinants in percent (map units).
Divide the plaque titer from su W225 by that from su⁺ W319 and
then multiply by 2, because only the (++), and not the (sus sus)
recombinants could be determined. That is, it is assumed that
the reciprocal recombinants occur with the same frequency. Determine the sequence of the sus markers x, y and z by comparing
the frequencies of recombinants in the two-factor crosses.

     Gene-order (map): ————————————————————————

Verify this order with the three-factor cross xz × y. The
basic assumptions are:

- relative distances of genetic markers are approximately
  additive,

- recombinants, which can be formed by 1 crossover, are more
  frequent than those which require 2 crossovers.

3. The average multiplicities of infection, m, i.e., the average
number of phages infecting a bacterium. For this, divide the
plaque titer of the parental phage-mixture by the cell titer
of the log culture of su⁺ W319 used for the cross (see Material).
This calculation is only valid if >90% of the parental phages
adsorbed to the bacteria. This is been the case, as shown in
previous experiments.

4. The average number of phage progeny per infected bacterium
(burst size). Divide the plaque titer on su⁺ W319 by the number
of bacteria in the original cross mixture (see Material). Keep
in mind that the bacteria have been diluted 1:2 when they were
added to 0.5 ml of the parental phage suspension. This calculation of burst size is only valid if the average multiplicity of
infection is ≫3 and more than 90% of the parental phages are
adsorbed.

Further evaluation possibilities:

5. The genetic location of the unselected marker c1 ("clear
plaque")
Preliminary note: The phage parent sus z contains, in addition
to the sus marker, one more mutation in the c1-gene; therefore,
it forms clear plaques, while those of the wild type are turbid.
a) Determine the frequency (in %) of the clear plaques among
all the plaques on su⁺ W319 (plates 7 + 8 or 11 + 12, respectively). (Values near 50% are expected, because the phage parents were
mixed at a ratio of 1:1 before the cross and none of them should
have had a selective advantage or disadvantage during the intracellular multiplication).

b) Calculate the frequency (in %) of the clear plaques among all
the plaques on su W225 (plates 5 + 6, 9 + 10, resp.). With these

values determine the position of the c1 gene in respect to x, y and z (see paragraph 2).

6. The <u>coincidence factor</u> for the cross xz × y:

$$C = \frac{F_{double}}{F_I \times F_{II}} = \underline{\hspace{2cm}} = \underline{\underline{\hspace{2cm}}}.$$

For this calculation first divide the map units determined in paragraph 2 by 100.

7. The <u>frequency of mixedly infected bacteria</u> in the crosses. The average multiplicity of infection calculated in paragraph 3 is composed of equal values m_1 and m_2, respectively, for each phage parent. The frequency of mixedly infected bacteria will be calculated according to POISSON (compare Expt. 12):

$$P = (1 - e^{-m_1}) \times (1 - e^{-m_2}) = \underline{\underline{\hspace{2cm}}}.$$

e = the natural base of logarithms.

Literature

DOERMANN, A.H., PARMA, D.H.: Recombination in Bacteriophage T4. J. Cell Physiol. <u>70</u>, Suppl. I, 147-164 (1967).

MOSIG, G.: Recombination in Bacteriophage T4. Advan. Genet. <u>15</u>, 1-53 (1970).

WINKLER, U., KOPP-SCHOLZ, U., HAUX, Ch.: Nonsense Mutants of Serratia Phage Kappa. Molec. Gen. Genetics <u>106</u>, 239-253 (1970).

<u>Time requirement</u>: 1st day 4 hrs, 2nd day 3 hrs.

Data sheet II (Evaluation)

Cross	Question (1) Plaque titer W225 (N_1) W319 (N_2)	Question (2) Map units (%) $\dfrac{N_1 \times 2 \times 100}{N_2}$	Question (3) Multiplicity of infection m	Question (4) Burst size	Question (5) Frequency of clear plaques on W319 W225
x × y					— —
y × z					
x × z					
xz × y					— —

18. Transduction of Linked and Unlinked Genetic Markers with Phage P1

A lysate of the temperate *E. coli* phage P1 contains, besides plaque-forming phages some "defective" phages. In these defective phage particles the phage DNA is replaced by a fragment of the chromosome of their last bacterial host. When a defective phage is adsorbed to a bacterium, it injects this DNA which then can recombine with the chromosome of the new host. Such a transfer of genetic material from one bacterial cell (donor) to another one (recipient) is called transduction. P1-phages can transduce any gene of a bacterium (<u>generalized transduction</u>). The frequency of transduction of a gene is about 10^{-5} to 10^{-6} per phage particle. If two genes are situated relatively close to each other on the bacterial chromosome, i.e., if the markers are closely linked, they can be transduced together by the same phage particle (<u>cotransduction</u>). Comparisons between genetic crosses by conjugation and by transduction show that the frequency of cotransduction of two genes is decreased with increasing distance from each other. Therefore it is usual to map closely linked genetic markers by transduction. If the distance between two genes is more than 2% of the total *E. coli* chromosome, they are no longer cotransducible, because DNA pieces of this length do not fit into the protein coat of phage P1. Non-cotransducible genes are often called "unlinked" genes, even if they are actually part of the same linkage group (chromosome).

In contrast to phage P1, phage λ transduces only genes next to its prophage insertion site on the *E. coli* chromosome, e.g., the genes <u>gal</u> or <u>bio</u> (<u>specialized transduction</u>). This difference is explained by the difference in the mode of formation of transducing P1 and λphages: transducing P1 is formed during its lytic multiplication; once in a while, instead of the phage DNA, some segment of the fragmented bacterial chromosome will be encapsulated by the phage protein. Transducing λ, however, can only be obtained after induction of lysogenic cells. Here, the prophage is not always excised as a whole; instead, only a part of the phage DNA together with a neighboring piece of the host chromosome will be encapsulated.

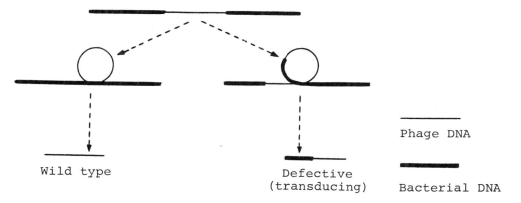

Wild type

Defective
(transducing)

Phage DNA

Bacterial DNA

A transducing chromosome fragment does not have to be recombined into the recipient cell's chromosome immediately after it has been injected. It can express its genetic information as a free piece of DNA (abortive transduction). During cell division this fragment will be linearly transferred, i.e., only one of the two daughter cells will receive the non-replicating piece of DNA. After several cell divisions, the fragment eventually can be incorporated into the cell's chromosome, so that genetically stable transduced cells are formed. Generally, in transduction experiments, multiplicities of infection smaller than 1 are used in order to reduce the possibilitiy that cells which were infected by a transducing phage, would later be lysed as a result of an additional infection with a non-defective phage.

Plan. Cells of recipient E. coli AB1157 (leu pro ara) will be infected with P1 phages previously grown on wild type (donor) bacteria. These will then be spread on two different selective media. On leucine and glucose-containing minimal agar, only those cells into which at least the selected pro^+ marker was transduced will form colonies. On proline and glucose-containing minimal agar, the transduction of the selected leu^+ marker will be correspondingly determined. By transferring these leu^+ and pro^+ colonies to other media, e.g. arabinose-EMB agar, it can be determined which of the three markers leu^+, pro^+ and ara^+ are cotransducible. For this test, replica plates will be made, using sterile velvets.

Material. 5 ml suspension of stat cells of E. coli AB1157 in 0.05 M Tris-HCl buffer, pH 7.9, with 0.002 M Ca^{++} (cell titer 5 × 10^8/ml). AB1157 carries the genetic markers pro, leu, thr, arg, his, thi, str^r, ara, xyl, mtl, lac, gal (see Appendix B). 1 ml suspension of phage P1 in the same Tris buffer (titer 2 × 10^9/ml). 1 round wooden block ("stamp") having the inner diameter of a petri dish and 6 sterile velvet pads for replica plating (see Expt. 11). 200 ml of Tris-HCl buffer as above. 2 TBY soft agar tubes.

    8 plates with TBY agar

    5 plates with M9 agar and 50 µg/ml proline (M9-P)

    5 plates with M9 agar and 50 µg/ml leucine (M9-L)

    6 plates with M9 agar (needed on 3rd day of the expt.)

    6 plates with eosine-methylene blue (EMB) agar containing
            5 mg/ml arabinose (needed on 3rd day of the expt.).

All 16 M9 (minimal agar) plates contain 50 µg/ml threonine, histidine, and arginine and 0.5 µg/ml thiamine, because the recipient AB1157 has more genetic markers than were necessary for this experiment.

Procedure

1st day (see data sheet I):
Titer of P1 phages. Dilute the phage suspension to 10^{-6} in Tris-HCl buffer. Pipette 0.1 ml each of the last dilution and 0.2 ml

each of the undiluted *E. coli* suspension into two small test
tubes, mix, and incubate at 37°C for pre-adsorption of the phages
to the bacteria. After 20 min add 3 ml of TBY soft agar and pour
the mixtures onto two TBY agar plates (Nos. 1 + 2).

Is the phage suspension free of bacteria? Spread 0.1 ml each
from the undiluted phage suspension onto two TBY agar plates
(Nos. 3 + 4).

Titer of the recipient cells. Dilute the *E. coli* suspension in
Tris-HCl buffer to 3×10^{-6}. Spread 0.1 ml each from the last
dilution onto two TBY agar plates (Nos. 5 + 6).

Titer of spontaneous back-mutants among the recipient cells.
Spread 0.1 ml each from the undiluted cell suspension onto two
M9-P agar plates (Nos. 7 + 8) and two M9-L agar plates (Nos. 9
+ 10).

Transduction. Mix 0.2 ml of the undiluted phage suspension with
1.8 ml of the undiluted *E. coli* suspension in a small test tube
and incubate at 37°C for 20 min. Then spread 0.1 ml each of
this pre-adsorption mixture onto 3 M9-P agar plates (Nos. 11 -
13) and 3 M9-L agar plates (Nos. 14-16).

Titer of surviving cells. Immediately after the end of the pre-
adsorption period, dilute the transduction mixture in Tris-HCl
buffer to 10^{-5}. Spread 0.1 ml each from the last dilution onto
two TBY agar plates (Nos. 17 + 18).

Incubate all plates at 37°C.

2nd day:
Count plaques and colonies, respectively, on the TBY agar plates
Nos. 1 - 6 and 17 - 18 and list results in data sheet I.

3rd day:
Count colonies on agar plates Nos. 7 - 16 and list results in
data sheet I. First, replica plate colonies from plates 11 - 16
with a sterile velvet stamp (see Expt. 11) to M9 agar plates
Nos. 11a - 16a and then to EMB agar plates Nos. 11 - 16 (data
sheet II). On the M9 agar plates only those colonies will grow
whose bacteria are leu⁺ and pro⁺ simultaneously. On EMB agar
all colonies will grow; those colonies from arabinose-fermenting
(ara⁺) cells have a blue-black color and show a golden-metallic
gloss, while colonies from ara cells appear white to pink.

Evaluation (4th day):

Count the colonies on the M9 and EMB agar plates and calculate
all the titers. The abbreviations used below are explained in
data sheet I.

1. The multiplicity of infection in the transduction mixture
will be calculated according to:

$$m = \frac{0.2 \times Ph}{1.8 \times R} = \underline{\qquad\qquad} .$$

2. The expected frequency of non-infected bacteria P_0 according to POISSON (see Expt. 12) is to be calculated; here it is assumed that 100% of the phages were adsorbed.

$$P_0 = e^{-m} = \underline{\qquad\qquad} .$$

(e = 2.72; natural base of logarithms)

From this we obtain the expected titer of the non-infected bacteria:

$$P_0 \times 0.9 \times R = \underline{\qquad\qquad} .$$

Does this theoretical titer conform with the titer of survivors (S) found in the experiment?

3. The effective frequency of transducing phage particles (transduction rate) will be calculated:

$$\frac{L - S_L}{0.1 \times Ph} = \underline{\qquad\qquad} = \underline{\qquad\qquad} . \qquad \text{(Transduction of } \underline{leu}^+)$$

$$\frac{P - S_P}{0.1 \times Ph} = \underline{\qquad\qquad} = \underline{\qquad\qquad} \qquad \text{(Transduction of } \underline{pro}^+)$$

4. Frequency of cotransduction
Cotransduction of $\underline{pro}^+$ with $\underline{leu}^+$:

$$\frac{\text{Total Colonies on 11a-13a}}{\text{Total Colonies on 11-13}} = \underline{\qquad\qquad} = \underline{\qquad\qquad}$$

Cotransduction of $\underline{leu}^+$ with $\underline{pro}^+$:

$$\frac{\text{Total Colonies on 14a-16a}}{\text{Total Colonies on 14-16}} = \underline{\qquad\qquad} = \underline{\qquad\qquad}$$

Data sheet I

Plate No.	Agar+Dilution of samples	Colonies or plaques	Colony- or plaque titer	Abbre- viations	Notes
1	TBY 10^{-6}			Ph	Titer of phase sus- pension
2	TBY 10^{-6}				
3	TBY undil.			–	Bacterial contam- ination in phage suspension
4	TBY undil.				
5	TBY 3×10^{-6}			R	Titer of recipient cells
6	TBY 3×10^{-6}				
7	M9-P undil.			S_L	Titer of spontaneous $\underline{leu}^+$ back mutants
8	M9-P undil.				
9	M9-L undil.			S_P	Titer of spontaneous $\underline{pro}^+$ back mutants
10	M9-L undil.				
11	M9-P undil.			L	Titer of cells transduced to $\underline{leu}^+$ (includ- ing back mutants)
12	M9-P undil.				
13	M9-P undil.				
14	M9-L undil.			P	Titer of cells transduced to $\underline{pro}^+$ (includ- ing back mutants)
15	M9-L undil.				
16	M9-L undil.				
17	TBY 1.1×10^{-5}			S	Titer of infect- ed recipient cells (sur- vivors)
18	TBY 1.1×10^{-5}				
11a	M9 undil.			–	Titer of co- transductants ($\underline{leu}^+$ and $\underline{pro}^+$)
12a	M9 undil.				
13a	M9 undil.				
14a	M9 undil.				
15a	M9 undil.				
16a	M9 undil.				

Data sheet II: Cotransduction of $\underline{ara}^+$ with $\underline{leu}^+$ and $\underline{pro}^+$, respectively

Plate No.	Evaluable colonies per plate blue-black	white-pink	Frequency of cotransduction
EMB 11			$\dfrac{\Sigma_b \times 100}{\Sigma_b + \Sigma_w} =$
EMB 12			
EMB 13			
	$\Sigma_b =$	$\Sigma_w =$	$\underline{}$ % $\underline{leu}^+$ $\underline{ara}^+$
EMB 14			$\dfrac{\Sigma_b \times 100}{\Sigma_b + \Sigma_w} =$
EMB 15			
EMB 16			
	$\Sigma_b =$	$\Sigma_w =$	$\underline{}$ % $\underline{pro}^+$ $\underline{ara}^+$

Which genes are closely linked with each other?

Literature

CARO, L., BERG, C.M.: P1 Transduction. Methods in Enzymology 21D, 444-458 (1971).

OZEKI, H., IKEDA, H.: Transduction Mechanisms. Annual Review of Genetics 2, 245-278 (1968).

TAYLOR, A.L., TROTTER, C.D.: Revised Linkage Map of *E. coli*. Bacteriological Reviews 31, 332-353 (1967).

Time requirement: 1st day 3 hrs, 2nd day 0.5 hrs, 3rd day 1 hr, 4th day 2 hrs.

19. Transfer and Elimination of a Plasmid (F'lac)

DNA molecules which can autonomously replicate in the bacterial
cytoplasm independent of the bacterial chromosome are called
"plasmids". While some of these plasmids are integrated into
the bacterial chromosomes at times (e.g., F factor; see Expt.
16), others probably always remain free in the cytoplasm (e.g.,
bacteriocinogenic factors; see Expt. 20). Plasmids are double-
stranded, covalently closed circular DNA molecules. Many plasmids commonly contain the genetic information for sex pili,
thread-like filaments on the bacterial surface, they can be
serologically differentiated into F- and I-pili. If bacteria
with sex pili by chance touch other cells without pili, the
cells maintain physical contact for a brief time, making possible the exchange of genetic material from one cell to the
other. Some "small" RNA and DNA phages (male specific phages)
specifically adsorb to the sex pili when they infect a cell.

In individual Hfr strains of *E. coli*, the F factor can be integrated into the bacterial chromosome at different sites. Changing
back into the cytoplasmic state, is a process similar to that occurring in the formation of transducing λdgal phages in λ-lyso-
genic bacteria (see figure in Expt. 18): Bacterial genes which
are located close to the insertion site of the F factor will be
excised together with episomal DNA: a so-called F' (F prime)
factor is formed. If, for example, the F prime factor contains
genes of the lac operon, it will be called F'lac. If the F'lac
cells are mixed with lactose-non-fermenting (lac) F⁻ cells, a
total plasmid will be transferred to the "recipient" cells making the recipient lac⁺. This process is called F-duction or
sex-duction.

How can F'lac cells be differentiated from Hfr cells which have
their origin (see Expt. 16) near the lac operon?

a) F'lac⁺ and other plasmids are eliminated if the corresponding
bacteria are cultured in broth of pH 7.6 to which acridine orange
has been added. Chromosomally integrated plasmids will not be
eliminated by this treatment.

b) All F'lac⁺ clones which have been isolated independently
from one another transduce their plasmids to other cells within
a few minutes. Different lac⁺ Hfr strains, however, transfer the
lac operon earlier or later, depending on the site of the respective origins.

c) If recipient cells have become lac⁺ by sex-duction, they always contain the F-factor. But recipient cells which have received the lac operon by conjugation with Hfr cells, possess
the F factor only when they have accepted the complete donor
chromosome.

Numerous bacterial strains have been isolated whose F factors
contain parts of the *E. coli* chromosome varying in size and gene

content. With these plasmids, stable partially diploid bacteria can be obtained which allow tests to be made on intergenic complementation (see Expt. 23) and on the dominance of certain alleles.

Plan. (1) Plasmid elimination. Two lac⁺ strains of *E. coli* will be separately cultured in broth containing acridine orange (pH 7.6). After about 10 generations, the cells will be spread on EMB lactose agar, in order to differentiate between lac⁺ and lac colonies. Acridine orange will cause the segregation of lac cells only from bacteria with extrachromosomal (plasmid) lac⁺ genes. (2) Plasmid transfer. Donor cells containing F'lac⁺ will be mixed with lac recipient cells which are also streptomycin resistant (strʳ). 15 min later this mixture will be plated on EMB lactose agar to which streptomycin has been added. Thus only strʳ recipient colonies can grow. The frequency of lac⁺ colonies on this indicator agar is a measure of the efficiency of plasmid transfer.

Material. 5 ml each of stat cultures in NB (37°C, aerated, in an Erlenmeyer flask) of: *E. coli* W1023 strˢ thi (lac pro)deletion F'lac⁺pro⁺, Hfr H3000 lac⁺ and W1022 lac (recipient). 50 ml each of NB with and without 20 µg/ml acridine orange (the NB was adjusted to pH 7.6 with 1 N NaOH, the acridine orange was dissolved in sterile NB (1 mg/ml); 1 ml of this solution was added to 49 ml of NB at pH 7.6). 12 EMB lactose agar plates. 9 EMB lactose agar plates containing an additional 20 µg/ml streptomycin. 1 petri dish with alcohol to sterilize the glass rod. 2 100-ml Erlenmeyer flasks with 9.9 ml TBY each. 100 ml NB. For several groups together: 1 vortex mixer. 1 water bath at 37°C.

Procedure (1st day)

1. Plasmid elimination (curing). Dilute stat cultures of *E. coli* W1023 and Hfr H3000 (control) 10^{-3} in NB and pipette from this 0.2 ml into each of the test tubes Nos. 1-4. Each of these 4 test tubes contains 10 ml NB, pH 7.6, with or without acridine orange (see data sheet I). Aerate the cultures 10-14 hrs at 37°C, until the cell titer approaches about 1×10^8/ml. Then dilute in buffer to 10^{-5} and spread 0.1 ml of the samples on EMB lactose agar. Incubate the plates at 37°C for about 20 hrs. Then count the dark-violet (lac⁺) and the pink-colored (lac) colonies separately and list in data sheet I.

2. Plasmid-transfer. Pipette 0.2 ml each of stat cultures of *E. coli* W1023 (F'lac⁺ donor) and W1029 (F'recipient) to 9.8 ml of TBY in an Erlenmeyer flask (100 ml capacity) and leave at 37°C until a cell titer of 4×10^8/ml is reached. Then proceed as follows:

A: 1 ml W1023 + 1 ml W1029

B: 1 ml W1023 + 1 ml TBY   = control

C: 1 ml TBY   + 1 ml W1029 = control

Incubate all three mixtures at 37°C for 15 min, dilute to 10^{-2} in NB (0.1 + 9.9 ml) and vigorously shake on the vortex mixer for 20 sec, in order to stop contact between the donor and the recipient cells. Dilute the mixtures with NB to 5×10^{-4} and spread 0.1 ml of the samples on EMB lactose streptomycin agar (data sheet II). Incubate the agar plates at 37°C for about 20 hrs. Count the dark violet (<u>lac</u>$^+$) and the pink-coloured (<u>lac</u>) colonies separately and list in data sheet II.

Data sheet I

Test tube No.	Strain	Acridine Orange µg/ml	Plate No.	Colonies Violet	Pink	Sum
1	W1023	–	1			
	W1023	–	2			
	W1023	–	3			
2	W1023	20	4			
	W1023	20	5			
	W1023	20	6			
3	Hfr H3000	–	7			
	Hfr H3000	–	8			
	Hfr H3000	–	9			
4	Hfr H3000	20	10			
	Hfr H3000	20	11			
	Hfr H3000	20	12			

Data sheet II

Mixture	Final dilution	Plate No.	Colonies Violet	Pink	Sum
A W1023 + W1029	5×10^{-6}	13			
	5×10^{-6}	14			
	5×10^{-6}	15			
B W1023 + TBY	5×10^{-6}	16			
	5×10^{-6}	17			
	5×10^{-6}	18			
C W1029 + TBY	5×10^{-6}	19			
	5×10^{-6}	20			
	5×10^{-6}	21			

Evaluation (2nd day)

1. Plasmid elimination (curing) (data sheet I). Separately calculate the frequency of the lac colonies (= pink colonies among
all the colonies) from the four NB cultures with and without
acridine orange.

Strain	Localization of lac$^+$	AO* μg/ml	Frequency of lac colonies
W1023	plasmid	-	
W1023	plasmid	20	
Hfr H3000	chromosome	-	
Hfr H3000	chromosome	20	

* AO: Acridine orange

2. Plasmid transfer (data sheet II). Calculate separately for
all three mixtures A, B, and C the frequency of the lac$^+$ colonies
(= violet colonies) among all colonies present.

Mixture	Frequency of lac$^+$ colonies
A	
B	
C	

3. Question. How could it be easily proven that the lac$^+$ recipient cells which have been formed by F-duction received the F
factor together with the lac$^+$ gene?

Literature

CAMPBELL, A.M.: Episomes. London: Harper & Row 1969.

HELINSKI, D.R.: Plasmid Determined Resistance to Antibiotics:
    Molecular Properties of R Factors. Ann. Rev. Microbiol. 27,
    437-470 (1973).

HIROTA, Y.: The Effect of Acridine Dyes on Mating Type Factors
    in *Escherichia coli* . Proc. Nat. Acad. Sci. US 46, 57-64 (1960).

LOW, K.B.: *Escherichia coli* K12 F-Prime Factors, Old and New.
    Bacterial Rev. 36, 587-607 (1972).

MEYNELL, E., MEYNELL, G.G., DATTA, N.: Phylogenetic Relationship
    of Drug Resistance Factors and Other Transmissible Bacterial
    Plasmids. Bacteriol. Rev. 32, 55-84 (1968).

SCAIFE, J.: Episomes. Ann. Rev. Microbiol. 21, 601-638 (1967).

Time requirement: 1st day 14 hrs (with an interruption of
                  several hrs), 2nd day 1 hr.

20. Induction of Lysogenic and Bacteriocinogenic Bacteria

<u>Lysogeny</u>. If bacteria are infected with temperate phages, the
cells will be either lysed after intracellular phage multiplication <u>or</u> they will become "lysogenic", i.e., they will survive with the unreplicated phage DNA entering the prophage
state. The prophage, which always consists of double-stranded
DNA, can either be integrated into the chromosome of the host
bacterium (e.g., phage λ) or it can exist as a plasmid (e.g.,
phage P1). Prophages will multiply synchronously with the bacterial DNA and will be passed on to both daughter cells during
cell division. The prophage state is maintained by a protein, the
repressor, coded for by the phage genome. The repressor molecules
prevent the transcription of those prophage genes which are necessary for the vegetative (= lytic) phage multiplication. The
repressor also causes lysogenic cells to show "immunity" to infection by homologous phages.

A prophage can spontaneously switch to the vegetative phase.
The chances for such a change are about 10^{-5} to 10^{-3} per cell.
This switch results in the lysis of the particular bacterium
and the release of infectious phage particles (spontaneous
induction). If lysogenic cells are treated with certain chemicals (nitrogen mustard, base analogues, mitomycin C, and others)
or are irradiated with UV light, more than 90% of the cells will
be induced simultaneously. Prophages which form a temperature-
sensitive repressor as a result of a mutation, can be induced
by raising the incubation temperature.

Lysogenization and induction are recombinational events and
are mediated by several enzymes (see CAMPBELL model in first
Reference). Experimental results suggest that certain animal
viruses can also exist in a state similar to that of the prophage.

<u>Bacteriocinogenic</u> bacteria contain, in the cytoplasm, circular
double-stranded DNA molecules (bacteriocinogenic factors), which
contain the information for the synthesis of a "bacteriocin".
Bacteriocins are antibiotics. For *E. coli*, more than 10 strains
are known which can produce different bacteriocins, called coli-
cins. Some *Bacillus*, *Streptococcus* and *Vibrio* strains are also
known to produce bacteriocins. Similar to temperate phages,
some bacteriocins are synthesized only after treatment of the
corresponding bacteria with UV light, mitomycin C or similar
chemicals; these cells are killed as a consequence of bacteriocin production (lethal synthesis). Other bacteriocins are not
inducible and the cells maintain their ability to divide in
spite of their bacteriocin production.

Bacteriocins probably consist only of protein. Their molecular
weights vary between 0.5 and 1×10^5 daltons. They inactivate
cells by adsorbing first to specific receptors on sensitive cells.
Then, depending on the type of bacteriocin, they, directly or indirectly, interfer with different cell functions, such as protein

synthesis (e.g., colicin E_3), DNA synthesis (e.g., colicin E_2) or energy metabolism (e.g., colicin E_1).

Some bacteriocinogenic factors code for bacteriocins and also for the formation of sex pili. These factors are transferable from bacterium to bacterium by cell contact (see F'lac transfer in Expt. 19). Bacteriocinogenic factors, antibiotic-resistant transfer factors and certain prophages are termed "plasmids".

Plan. λ-lysogenic E. coli bacteria will be irradiated with UV. The survivors, the percentage of induced cells and the number of released λ-phages per induced bacterium, will be determined. Similarly colicinogenic bacteria will be irradiated; the survival of the cells and the increase of the extracellular colicin concentration as a consequence of colicin induction will be determined. As controls, non-lysogenic and non-colicinogenic bacteria will be irradiated to determine the lethal effect of UV, independent of the lethality due to the induction.

Material. 15 ml log cultures in NB ($37^{O}C$, aerated) of E. coli strains

K12(λ)λ^r: lysogenic and resistant for phage λ (the resistance prevents readsorption of progeny phages).

W3110 col D: colicinogenic for colicin D.

K12s: non-lysogenic and non-colicinogenic indicator for λ phages.

Cell titer 3×10^8/ml. 5 ml stat culture of E. coli K12s nal^r, a nalidixic acid-resistant indicator for colicin D, which has been diluted to 2×10^7 cells/ml in P-buffer. 20 NB agar plates. 8 NB soft agar tubes. 10 12-ml centrifuge tubes. 3 sterile glass petri dishes. 50 ml NB. 1 ml nalidixic acid, 800 µg/ml. Glass dish with alcohol for flaming the glass rod. Polystyrene bucket with ice. For several groups together: 1 ozone-free Hg low pressure (ultraviolet) lamp, horizontally mounted, in a metal housing with a sliding shutter, photon flux-rate about 20 erg mm^{-2} sec^{-1}. (For methods of standardization, see pp. 206 and 207.) 1 small centrifuge. Yellow light.

Procedure (1st day)

Carry out the following work under yellow light in order to avoid undesired photoreactivation. Wear protective goggles during the UV irradiation. Take experimental details from data sheet I.

Viable count of non-irradiated bacteria. Centrifuge 10 ml each of log cultures K12s, K12(λ)λ^r and W3110 colD at about 4,500 rpm for 10 min. Discard the supernatant and resuspend the pellet in 10 ml of ice cold P-buffer. Dilute the suspensions and spread plates Nos. 1-6 with 0.1 ml of the desired dilution.

UV irradiation; viable count of the survivors. Irradiate 5 ml each of the above suspensions in petri dishes for 20 sec with

with UV light, dilute, and spread 0.1 ml each of the desired dilution on plates Nos. 7-12. In addition, prepare soft agar layer plates Nos. 13 and 14 from the suspension of lysogenic cells to determine the titer of the induced cells. Use 0.1 ml diluted cell suspension, 0.2 ml indicator K12s and 3 ml soft agar per plate.

Phage and colicin production. Centrifuge the non-irradiated and the irradiated suspensions of K12(λ)λ^r and W3110 colD again at 4,500 rpm for 10 min, discard supernatants and resuspend the pellets in 5 ml NB each. Aerate these suspensions at 37°C for 2 hrs. From here on the work can be done in daylight. Centrifuge all four cultures at 4,500 rpm for 10 min, decant and retain the supernatants.

Titers of λ phages. Add 5 drops of chloroform to each supernatant of the two lysogenic cultures, shake for a short time by hand, and then dilute and start soft agar layer plates Nos. 15-18. Before the addition of the soft agar, allow the phages to adsorb to the indicator bacteria, K12s, at 37°C for 15 min.

Titers of colicin D (drop test). Start two soft agar layer plates (Nos. 19 and 20) with 0.1 ml each of stat culture of K12s nal^r. Add 0.25 ml nalidixic acid solution to the soft agar, to prevent the growth of surviving colicinogenic bacteria. Dilute the supernatants of both colicinogenic cultures in NB following data sheet II and plate each culture on one of the two plates. Beginning with the greatest dilution, place small drops on the plates in a clockwise direction using a 0.1 ml pipette. Use one pipette for each culture. The drops should contain about 0.01 ml. As a control, place 1 drop of NB in the middle of each of the plates.

Incubate all plates at 37°C overnight.

Data sheet I

Plate No.	UV sec	Strain	Dilution	Plate	Colonies or plaques	Titer	Reference letter
1	O	K12s	10^{-5}	Streak			A
2	O	"	"	"			
3	O	K12(λ)λ^r	"	"			B
4	O	"	"	"			
5	O	W3110 co1D	"	"			C
6	O	"	"	"			
7	20	K12s	10^{-5}	Streak			D
8	20	"	"	"			
9	20	K12(λ)λ^r	10^{-4}	"			E
10	20	"	10^{-5}	"			
11	20	W3110 co1D	10^{-4}	"			F
12	20	" "	10^{-5}	"			
13	20	K12(λ)λ^r	10^{-5}	Pour			G
14	20	"	"	"			
15[a]	O	"	10^{-4}	"			H
16[a]	O	"	10^{-5}	"			
17[a]	20	"	10^{-6}	"			I
18[a]	20	"	10^{-7}	"			

[a] Start these plates after the appropriate cultures have been incubated for 2 hrs.

Data sheet II

Plate No. 19: Supernatant from non-irradiated W3110 co1D

Dilution	Result[a]	Dilution	Result[a]
undil.		1 : 16	
1 : 2		1 : 32	
1 : 4		1 : 64	
1 : 8		1 : 128	

Plate No. 20: Supernatant from irradiated W3110 co1D

Dilution	Result[a]	Dilution	Result[a]
1 : 10		1 : 160	
1 : 20		1 : 320	
1 : 40		1 : 640	
1 : 80		1 : 1280	

[a] Complete lysis = +; Clearing = ±; Looks like control = -

Evaluation (2nd day)

Bacterial strain	Ratio	Survival
K12s	D/A =	
K12(λ)λ^r	E/B =	
W3110 colD	F/C =	

- Fraction of induced λ-lysogenic cells
  G/B =

- Average burst size of induced λ-lysogenic cells:
  (I-H)/G =

- Record the concentration of extracellular colicin D in relative units (RU). One RU/ml is that amount which is able to
  prevent the growth of sensitive indicator cells, under the
  test conditions chosen, so that a clear zone is formed at the
  site of the drop.
  RU, non-irradiated W3110 colD:
  RU_{UV}, irradiated W3110 colD:
  Colicin increase by induction: RU_{UV}/RU =

Literature

CAMPBELL, A.M.: Episomes. London: Harper & Row 1969.

DOVE, F.W.: The Genetics of the Lambdoid Phages. Ann. Rev.
  Genetics 2, 305-340 (1968).

ECHOLS, H.: Lysogeny: Viral Repression and Site-Specific Re-
  combinations. Ann. Rev. Biochem. 40, 827-854 (1971).

HELINSKI, D.R., CLEWELL, D.B.: Circular DNA. Ann. Rev. Biochem.
  40, 899-942 (1971).

REEVES, P.: The Bacteriocins. In: Molecular Biology, Biochemistry
  and Biophysics, Vol. 11. Berlin-Heidelberg-New York: Springer
  1972.

RICHMOND, M.H.: Resistance Factors and Their Ecological Importance
  to Bacteria and to Man. Progress in Nucleic Acid Research and
  Molecular Biology 13, 191-248. New York: Academic Press 1973.

Time requirement: 1st day 4.5 hrs, 2nd day 2 hrs.

21. Transfection of *E. coli* Spheroplasts with DNA of Phage Lambda

Transfection is spoken of if bacteria form infectious phage progeny after the uptake of free phage DNA. Usually two prerequisites are necessary for this: (1) The cells must be able to take up nucleic acids of high molecular weight (the cells must be competent; (2) one entire phage genome must enter the cell. Sometimes reconstitution of an entire genome takes place by recombination of overlapping fragments.

With certain bacterial strains (*Bacillus subtilis* and *Haemophilus influenzae*), competence is reached by special growth conditions. In *E. coli* K12, transfection is successful when the cell wall of the bacteria has been removed, e.g., enzymatically, before the addition of DNA. The spheroplasts which are formed by this treatment are held together by the cell membrane and are only stable in hypertonic medium.

The efficiency of transfection is the ratio of infectious phage progeny to the number of phage DNA molecules added in the transfection experiment. The efficiency of transfection of spheroplasts is $\leqslant 10^{-2}$.

Plan. Growing *E. coli* cells will be transformed to spheroplasts by the combined action of the chelating agent EDTA and the cell wall dissolving enzyme lysozyme. These spheroplasts will be transfected with free DNA, which was isolated from λ phages. The efficiency of transfection will be determined as a function of the DNA concentration used. Furthermore, it will be shown that the transfecting λ DNA is sensitive to pancreatic deoxyribonuclease (DNAse).

Material. 11 ml of a log culture of a λ-adsorption resistant mutant (λ^r) of *E. coli* AB1157 in TBY broth (aerated, 37°C). 5 ml of a log culture of λ-sensitive (λ^s) AB1157 for the titering of λ-phages. The cell titer of both cultures should be approx. 4×10^8/ml. 0.2 ml of a lysozyme solution with a concentration of 2 mg/ml. 2 ml of a 4% EDTA solution (Na_2-ethylenediamine tetraacetic acid. 1 ml of 0.1 M Tris-HCl buffer, pH 7.9. 1 ml of a 1.5 M sucrose solution. 200 ml of 0.05 M Tris-HCl buffer, pH 7.9, containing 4% sucrose. 20 ml of TBS (TBY broth containing 10% sucrose and 0.1% $MgSO_4 \times 7H_2O$). 1.2 ml TBSS (TBS containing 2% bovine serum albumin, electrophoretically purified). 1 ml protamine sulfate with a concentration of 1 mg/ml. 0.3 ml of a DNAse solution with a concentration of 0.7 mg/ml with 0.05 M $MgSO_4$. 0.5 ml of a λ DNA solution in Tris-HCl buffer, $O.D._{260}=0.3$. 1 small centrifuge. 2 10-ml centrifuge tubes. 1 small piece of absorbent paper. 18 TBY agar plates. 18 TBY soft agar tubes, 3 ml each. 37°C water bath. Vortex mixer. 2 sheets linear graph paper.

Procedure

Volumes used in the transfection assay (for exact amounts, see data sheet I):

1   vol DNA solution

1   vol Tris-HCl buffer

2.5 vol suspension of spheroplasts

30 min incubation at 37°C (uptake of DNA molecules), then addition of 6.5 vol TBS.

3 hrs incubation at 37°C (intracellular multiplication of phages); then titering of progeny phages.

1. Dilution of λ DNA solution. If possible, carry out all dilutions with 1 ml pipettes. In order to avoid damage to the DNA molecules by shearing forces, fill and empty the pipettes gently. Dilute the DNA solution to 10^{-1}, 10^{-2} and 10^{-3} in Tris-HCl buffer: 0.1 ml DNA + 0.9 ml buffer, etc. Mix the DNA with the buffer by rotating the tube in a sloping position. Label 6 small test tubes following data sheet I, then add first 0.2 ml of the corresponding DNA solutions and then the Tris-HCl buffer.

2. Preparation of spheroplasts. Centrifuge 10 ml of a log culture of AB1157 λ^r in a centrifuge tube at 2,000 rpm for 8 min. Then completely decant the supernatant and carefully dry the inside of the centrifuge tube with sterile absorbent paper. Add 0.05 ml of 0.1 M Tris-HCl buffer and 0.175 ml of a 1.5 M sucrose solution to the pellet. Resuspend the cells with a vortex mixer. At room temperature add the various solutions according to the following time table (stop watch!) and mix by gentle swirling:

t = 0   min: 0.018 ml lysozyme solution

t = 1   min: 0.010 ml EDTA solution

t = 9.5 min: 1.0   ml TBSS

t = 15   min: 3.5 ml TBS and then 0.25 ml protamine sulfate solution, mix.

t = 20   min: Check suspension microscopically (phase-contrast microscope, magn. 10 × 40). More than 90% of the originally moving rods should now be seen as non-moving, globular spheroplasts.

Spheroplasts are now ready for transfection.

3. Transfection. This begins with the addition of spheroplasts to the prepared DNA tubes following data sheet I.

Two controls will be included:

to (2b) pipette 0.1 ml DNAse solution 10 min prior to the addition of the spheroplasts.

to (2c) add 0.5 ml of a log culture of AB1157λ^r instead of the spheroplast solution.

Mix the tubes by gently swirling and then incubate in a 37°C water bath (uptake of DNA into the spheroplasts). After 30 min (end of uptake) pipette 1.3 ml TBS to each tube and continue incubation for 3 hrs so that phage progeny will be formed and can be released.

Data sheet I

Tube No.	Preparation		DNA uptake $t = 0$	$t = 30$ min
	ml DNA solution	ml buffer	ml suspension	ml TBS
1	0.2 undil.	+ 0.6	+ 0.5 Spheropl.	+ 1.3
2a	0.2 10^{-1}	+ 0.6	+ 0.5 Spheropl.	+ 1.3
2b	0.2 10^{-1}	+ 0.5[a]	+ 0.5 Spheropl.	+ 1.3
2c	0.2 10^{-1}	+ 0.6	+ 0.5 Cells	+ 1.3
3	0.2 10^{-2}	+ 0.6	+ 0.5 Spheropl.	+ 1.3
4	0.2 10^{-3}	+ 0.6	+ 0.5 Spheropl.	+ 1.3

[a] 10 min before the addition of the spheroplasts, add 0.1 ml of the DNAse solution.

4. Titers of phage progeny (with preadsorption). Dilutions and plates correspond to data sheet II. For preadsorption, pipette 0.1 ml of the phage suspension into a small tube, and add 0.2 ml of a log culture of AB1157. After mixing, incubate at 37°C for 15 min. Then add 3 ml soft agar to the tube and pour the contents onto a TBY plate.

Incubate plates at 37°C for about 20 hrs and then count plaques.

Evaluation

1. Proportionality of phage progeny and λ DNA concentration. Plot the sum of the λ phages per tube (Nos. 1, 2a, 3, 4) as a function of the DNA dilution in a double logarithmic scale. The slope of the curve should be interpreted. What can be concluded from the linearly increasing part and what is the meaning of the plateau?

2. Calculation of the transfection efficiency. Here the number of λ DNA molecules which were added to the transfection experiment must be known. It is calculated as follows: One λ DNA molecule weighs 7.7×10^{-17} g. A λ DNA solution with an $O.D._{260} = 1.0$ contains 50 µg DNA/ml (= 5×10^{-5} g/ml) and thus

$$\frac{5.0 \times 10^{-5}}{7.7 \times 10^{-17}} = 6.5 \times 10^{11} \text{ λ DNA molecules/ml.}$$

Data sheet II

Tube No.	Dilution	Plate No.	Plaques per plate	Titer	ΣPhages/Tube = titer $\times$ 2.6[a]
1	undiluted 10^{-1} 10^{-2} 10^{-3}	1 2 3 4			
2	undiluted 10^{-1} 10^{-2} 10^{-3}	5 6 7 8			
2b 2c	undiluted undiluted	9 10			
3	undiluted 10^{-1} 10^{-1} 10^{-2}	11 12 13 14			
4	undiluted 10^{-1} 10^{-1} 10^{-2}	15 16 17 18			

[a] The factor 2.6 results from the initial volume in the transfection tube.

Calculate the transfection efficiency for each of the concentrations used in the test (10^0 to 10^{-3}):

Tube No.	
1	$\dfrac{\Sigma\text{Phages/Tube}}{0.2 \times 0.3 \times (6.5 \times 10^{11}) \times 10^0} = \underline{\qquad} = \underline{\qquad}$
2a	$\dfrac{\Sigma\text{Phages/Tube}}{0.2 \times 0.3 \times (6.5 \times 10^{11}) \times 10^{-1}} = \underline{\qquad} = \underline{\qquad}$
3	$\dfrac{\Sigma\text{Phages/Tube}}{0.2 \times 0.3 \times (6.5 \times 10^{11}) \times 10^{-2}} = \underline{\qquad} = \underline{\qquad}$
4	$\dfrac{\Sigma\text{Phages/Tube}}{0.2 \times 0.3 \times (6.5 \times 10^{11}) \times 10^{-3}} = \underline{\qquad} = \underline{\qquad}$

3. The fraction of competent spheroplasts will be calculated as follows:

$$\frac{\Sigma\lambda \text{ phages (in the tube with the highest titer)}}{\Sigma \text{ the spheroplasts in tube} \times 250} = \underline{\hspace{2cm}}$$

It is assumed here that each transfected spheroplast produces an average of 250 λ phages (results of previous tests), and that the spheroplast titer is twice as high as the titer of the original cell suspension (10 ml cell suspension resulted in about 5 ml spheroplast suspension).

4. <u>Questions</u>

a) Why were λ-resistant cells used for the preparation of spheroplasts instead of λ sensitive cells?

b) Was transfection found after DNAse treatment of λ DNA? Yes/No; Comment:

<u>Literature</u>

BENZINGER, R., KLEBER, I., HUSKEY, R.: Transfection of *Escherichia coli* Spheroplasts I. J. Virology <u>7</u>, 640-650 (1971).

WACKERNAGEL, W.: An Improved Spheroplast Assay for λ-DNA and the Influence of the Bacterial Genotype on the Transfection Rate. Virology <u>48</u>, 94-108 (1972).

<u>Time requirement:</u> 1st day 6 hrs, 2nd day 1.5 hrs.

Problems

1. In two-factor crosses of the type a × b, b × c, a × c, etc., the gene sequence a, b, c,....x, y, z was determined. The sum of the relative individual distances between a and z was larger than a 50% recombination frequency (50 recombination units).

In a three-factor cross, ab^+z × a^+bz^+, a^+b^+ recombinants were selected. Among these, 65% got the allele z^+ and 35% the allele z. However, 50% for each was expected, because even and odd numbered exchanges are supposed to occur at an equal frequency if the distance between genes is large. How can this result be explained?

2. If bacteria are simultaneously infected with phages carrying different genetic markers and if later in phage development the progeny DNA is isolated from these cells, DNA molecules can be detected which carry DNA of both parental genotypes. Early in phage development the parental DNA components are joined together solely by hydrogen bonds (joint molecules); later, they are co-valently linked (recombinant molecules). The same has also been found for bacterial conjugation. How should the DNA of both parents be labeled and which methods should be employed to differentiate between joint and recombinant molecules?

E. Phenotypic Expression

The phenotypic expression of DNA begins with transcription. The
genetic information which is thus passed on to mRNA is then
translated into amino acid sequences. A large number of proteins
result from this process. These proteins are either components
of cellular structures or have catalytic or regulatory functions
in cell metabolism.

Some aspects of transcription have already been discussed in
Section II, since the formation of specific mRNA can only be
proven by the methods of nucleic acid chemistry. In one of the
following experiments (No. 24), the substrate-induction of β-
galactosidase will be demonstrated as an example of a transcription control mechanism. This experiment is part of the in-
direct proof of the "negative control" of transcription. Positive
control of transcription, e.g., via the sigma factor of the RNA-
polymerase, is experimentally more elaborate to show.

Translation, i.e., the actual biosynthesis of proteins, is not
dealt with here, because a book on experimental methods concerning this subject has just recently been published (Last and
Laskin). Part of this field are the isolation and characterization of ribosomes, of the "soluble fraction" (100.000 xg supernatant), and of the products of protein biosynthesis. This also
includes the study of the mode of action of certain antibiotics
which either cause miscoding (e.g. streptomycin) or completely
inhibit protein synthesis.

In genetic studies of intermediary metabolism, we are especially
interested in the number of genes which take part, either directly or indirectly, in a certain reaction or in the formation of
a certain reaction product. A method often applied to the solution of this problem is the Cis-trans test, which will be introduced as a complementation test with phages in Expt. 23. Another
helpful method, is synthrophy or cross-feeding (Expt. 22). With
this test, a biosynthetic pathway can formally be subdivided
into a series of single reactions which are under the control
of different genes.

Literature

DICKERSON, R.E., GEIS, I.: The Structure and Action of Proteins.
  New York: Benjamin 1969.

LAST, J.A., LASKIN, A.I.: Protein Biosynthesis in Bacterial
   Systems. Methods in Molecular Biology 1, 333 pp. New York:
   Dekker Inc. 1971.

MANDELSTAM, J., McQUILLEN, K. (eds.): Biochemistry of Bacterial
   Growth, 540 pp. Oxford: Blackwell Scientific 1968.

SCHLESSINGER, D.: Genetic and Antibiotic Modification of
   Protein Synthesis. Ann. Rev. Genetics 8, 135-154 (1974).

22. Determination of a Metabolic Block by Auxanography and Syntrophy

Many bacteria are <u>heterotrophic organisms,</u> i.e., they need organic substances, in order to satisfy their carbon and energy requirements. Heterotrophic bacteria can be subdivided into:

- <u>Prototrophic bacteria</u>. They grow in minimal medium, e.g., a solution which consists of only inorganic salts and glucose. These bacteria synthesize all cell components by themselves.

- <u>Auxotrophic bacteria</u>. These bacteria only grow in minimal medium which has been supplemented with defined growth factors, e.g., an amino acid, a vitamin or a purine base, or in broth, which is composed of a mixture of many growth factors. Many naturally occuring bacterial species are auxotrophic for one or more growth substances. Auxotrophic bacteria can, however, also originate from prototrophic bacteria by mutation. In such a case, usually only a single enzyme loses its function.

Many growth factors are synthesized in consecutive steps, catalyzed by different enzymes. Each enzyme is determined by at least a single gene (one gene-one polypeptide-hypothesis).

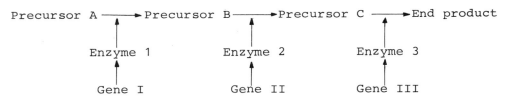

Therefore, mutations in gene I or II or III make the cell dependent on the same growth factor. It should be pointed out that substances not being regarded as growth factors are also synthesized in multi-step reactions, e.g., bacterial pigments or non-peptide antibiotics.

In the following, two procedures will be described by which a "genetic block" in a biosynthetic pathway can be investigated: auxanography and syntrophy. Both methods, to a certain extent, supplement one another.

<u>Auxanography</u> (explained with auxotrophic mutants as an example). Sterile filter discs, soaked with mixtures of different dissolved growth factors, will be placed on minimal agar. This agar was previously inoculated with a dense suspension of bacteria with an unknown growth factor requirement. When incubated, the cells multiply only in the region of diffusion around those filter discs which contain the necessary growth factor. The principle of auxanography is that each growth factor is present in only two different test solutions and that any pair of two test solutions always has only a single growth factor in common. In this way 36 different growth factors can be tested for their

suitability with only 9 test solutions. The calculation is as follows:

$$x = \frac{n!}{(n-2)! \times 2}$$

n = Number of test solutions
x = Maximum number of possible growth factors in n test
    solutions

Example: n = 4   Plan to mix test solutions:

$$x = \frac{1 \times 2 \times 3 \times 4}{(1 \times 2) \times 2} = 6$$

Test solution	Growth factors
A	1 + 2 + 3
B	1 + 4 + 5
C	2 + 4 + 6
D	3 + 5 + 6

Syntrophy (explained with mutants defective in pigment synthesis as an example). If an enzyme involved in the synthesis
of prodigiosin* loses its catalytic activity by mutation, the
substrate ("precursor") of the inactive enzyme will still continue to be synthesized. This results in an accumulation of the
precursor and its release into the medium. Another prodigiosin
defective mutant may utilize this precursor as long as its genetic block does not prevent the processing of the precursor to
prodigiosin (see scheme).

Mutant X
――――――――――――― E_1 E_2 E_3
(Enzyme 3 inactive)   A ――――――→ B ――――――→ C ――――//→ no pigment
                                                    |
                                                    |
                                                 Feeding
                                                    |
Mutant Y |
――――――――――――― |
(Enzyme 2 inactive)   A ――――――→ B ――//→ ↓――――――――→ pigment

This scheme shows that only mutant Y is fed by mutant X, but
not *vice versa*.

Plan. (1) The nature of the growth factor requirement of four
different auxotrophic mutants of *Serratia marcescens* will be determined by auxanography. (2) Three mutants of *Serratia marcescens* defective in pigment synthesis (pig) will be tested for
their ability to synthesize the dark red pigment (typical of
the wild type) under syntrophic conditions. The feeding pattern
found, will be used to determine which of several formal possibilities for the pathway of pigment biosynthesis holds true.

―――――――――――――

* Prodigiosin is the red pigment formed by *Serratia marcescens*
(see p. 149 for structural formula).

Material

Auxanography. 1 ml of stationary cells of auxotrophic mutants W366, W626, W861 and W982 of *Serratia marcescens* (cell titer approx. 1×10^7/ml). The suspensions were prepared in P-buffer with cells grown on NB agar at 30°C for about 24 hrs. 4 M9 agar plates. 4 soft agar tubes without NB (!). 1 sterile petri dish with about 20 filter discs (washed with distilled water, autoclaved and dried). The filter discs were prepared from filter paper with an office-punch. 1 pair of pointed forceps. 2 paper towels. 1 beaker with about 200 ml of tap water. Test solutions 0.5 ml containing each of A, B, C and D, which were made up by mixing equal volumes of the growth factors solutions Nos. 1-6, as required (see scheme on p. 145):

Growth factor 1: 3×10^{-2} M thymidine (MW 242)
solution
"   2: 3×10^{-5} M thiamine dichloride (MW 337)
"   3: 3×10^{-2} M L(-)-histidine (MW 155)
"   4: 3×10^{-2} M L-leucine (MW 131)
"   5: 3×10^{-2} M L(+)-lysine monohydrochloride
            (MW 183)
"   6: 3×10^{-2} M L(-)-methionine (MW 149)

Thus the final concentrations of the growth factors in test solutions A-D are 1×10^{-2} or 1×10^{-5} M.

Syntrophy. 3 ml each of a stat culture (in NB, grown at 30°C) of *Serratia marcescens* W225 (wild type) and the pig mutants W592, W622 and W623, adjusted to a cell titer of about 1×10^9/ml. 11 PG agar plates. About 20 cm of scotch tape.

Procedure (1st day)

1. Auxanography. Label four M9 agar plates with W366, W626, W861 and W982 respectively, and subdivide each plate into 4 sectors (A, B, C and D) of equal size. All markings should be on the bottom of the plate. Pour 0.1 ml of each bacterial suspension together with soft agar onto the correspondingly labeled agar plates. Allow the soft agar to solidify. With a pair of forceps, dip one filter disc into test solution A and make sure that no excess of solution is on the filter. Then place this disc on sector A of the M9 agar plate. Continue as before with new filter discs until all A sectors of the 4 agar plates have discs which have been soaked in solution A. Rinse the forceps with tap water and dry with a paper towel. Repeat the same procedure with the other test solutions B, C and D, respectively. Incubate agar plates at 30°C for 2 days.

2. Synthrophy
a) Control plates are prepared in order to recognize pigment forformation of the bacteria under non-synthrophic conditions. Label each PG agar plate with W225, W592, W622 and W623 and inoculate as follows: make two parallel streaks on the agar plates with the inoculation loop (Ill. 1 on p. 147).

Streak	PG plates		
a	225	592	225
b	622	622	592
c	623	623	-

Ill. 1   Ill. 2

b) <u>Syntrophic plates</u>. Streak the four bacterial suspensions with sterile inoculating loops onto 3 previously labeled PG agar plates (see Ill. 2 and the table). The streaks should almost touch one another (approx. 1 mm distance). Take care that the streaks do not run into each other!

c) <u>Test for volatility of a pigment precursor</u>. Mark two PG agar plates with W592 and streak the bacterial suspension over the entire surface with an inoculating loop (many zig-zag lines). Label each of 1 PG agar plate with W622 and W623 and inoculate corresponding to Ill. 1. Discard the lids to these 4 plates. Then put the W622 and the W592 agar plates together, with the agar sides facing each other, and loosely tape together with scotch tape. Do not make an "anaerobic chamber" because *Serratia* bacteria are obligatory aerobes. Repeat with W623 and the other W592 agar plate. Incubate all plates at 30°C for one day.

Evaluation

1. <u>Auxanography</u> (data sheet I; 3rd day). Which filter discs are surrounded by growth halos? Give approximate diameter and pigmentation of the halos. On which agar plates is heavy "background-growth" (this is the bacterial growth on the agar surface outside of the growth halo)? Discuss possible causes for the heavy background growth. The mutants are auxotrophic for which growth factors?

Data sheet I

Bacterial strain	Growth halo around filter				Auxotrophic for	Other observations (background, etc.)
	A	B	C	D		
W366 W626						
W861 W982						

2. Syntrophy (data sheet II, 2nd day). At first observe the pigmentation of the bacteria on the control plates and compare with those on the synthrophic plates and the "double-plates".

Data sheet II

Acceptor[a]	Color of inoculated streak on control plates	Inducer[a] 225 592	622 623
W225 W592			
W622 W623			

(-): no pigment-induction

(+) or (++): Weak, or heavy pigment induction

---

[a] An "inducer"-strain releases a precursor into the medium and an "acceptor"-strain can make use of it. Note that a strain may be inducer and acceptor simultaneously.

From the syntrophic behaviour of the three pig mutants, conclusions can be drawn as to which one of the four schemes given below for the biosynthesis of the pigment might be correct. Then write the number of the mutant on the reaction step corresponding to the location of the genetic block in that mutant. A, B, C = any hypothetical pigment precursor.

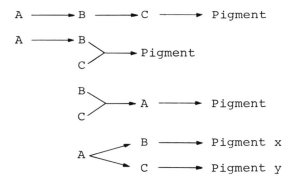

Questions

- Which one of the two mutants W622 and W623 can use a volatile pigment precursor?

- How can the syntrophic behaviour of the wild type W225 be explained?

- How can the following observation be explained: one of the
  pigment mutants does not form white or pale pink colonies,
  but rather orange-yellow ones?

Literature

GREENBERG, D.M. (ed.): Metabolic Pathways. London: Academic
Press. Especially Vol. III, Amino Acids and Tetrapyrroles
(1969); Vol. IV, Nucleic Acids, Protein Synthesis and Co-
enzymes (1970); Vol. V, Metabolic Regulation (ed. H.J. VOGEL)
(1971).

GUNSALUS, I.C., STANIER, R.Y. (eds.): The Bacteria, Vol. III.
Biosynthesis. London: Academic Press 1962.

MORRISON, D.A.: Prodigiosin Synthesis in Mutants of *Serratia
marcescens*. J. Bacteriol. $\underline{91}$, 1599-1604 (1966).

WILLIAMS, R.P., HEARN, W.R.: Prodigiosin. In: Antibiotics,
Vol. 2 (Biosynthesis), pp. 410-432, Addenda: pp. 449-450.
Berlin-Heidelberg-New York: Springer 1967.

Time requirement: 1st day 2 hrs, 2nd day 1 hr (syntrophy),
3rd day 1 hr (auxanography).

Prodigiosin

23. Complementation of Nonsense Mutants of Phage Kappa

Mutants which are phenotypically identical can be allelic or
non-allelic, i.e., they may have originated by mutation in homologous or non-homologous genes. Non-allelism indicates that
more than one gene product is necessary for the formation of a
particular character that distinguishes the wild type from the
mutant. By complementation tests it can be determined whether
two mutants are allelic or non-allelic. In the following complementation test nonsense (sus) mutants of a temperate phage
will be used (sus = suppressor sensitive).

Phage	su bacteria	su$^+$ bacteria
Wild type	+	+
sus mutant	-	+

A phage mutant with a nonsense triplet in an essential gene does
not produce infectious progeny in non-permissive (su) bacteria.
In cells of a permissive (su$^+$) strain the phenotypic expression
of nonsense triplets is suppressed. Sus-mutants can arise by
mutations in many different genes, and all of them have more
or less "identical" phenotypes, namely, lack of plaque formation on su indicator. If su bacteria are simultaneously infected
by two non-allelic sus mutants, "intercistronic complementation"
can occur, i.e., the physiological defect which is caused by the
mutation in one mutant is compensated for by the wild type allel
in the other mutant, and infectious phage progeny are produced
(Ill. 1). If, however, the phage mutants are allelic, the
mutual physiological complementation can not occur (Ill. 2).

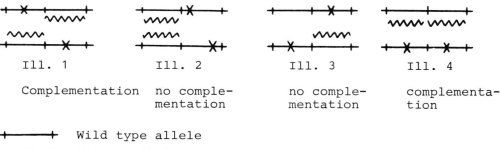

Ill. 1	Ill. 2	Ill. 3	Ill. 4
Complementation	no comple- mentation	no comple- mentation	complementa- tion

+———+   Wild type allele

+—✕—+   Mutant allele (recessive, except in the left gene
in Ill. 3)

〰〰〰   functionally active polypeptide as gene product

Generally complementation tests are carried out only with mutations in the *trans*-position, i.e. situated on different chromosomes (see scheme). Such an experiment is performed here and it
is tacitly assumed that the mutated alleles are recessive. If,

as in certain cases, this assumption does not hold true (e.g., some operator mutations in bacteria), then complementation does not work even if the mutations lie in non-allelic cistrons (Ill. 3, left gene). Therefore, a *trans* test (Ill. 1) should, always be accompanied by a *cis* test (Ill. 4), where the mutations are located on the same chromosome. Thus, *trans* tests showing no complementation only give evidence of allelism if complementation is found in the corresponding cis test.

If missense mutants of phages, e.g., temperature sensitive (<u>ts</u>) mutants, are tested for mutual complementation, some allelic mutants can simulate non-allelism by weak "intracistronic complementation". This phenomenon, however, is not found with nonsense mutants (why?).

For phage complementation studies bacteria are infected with two genetically different phage types. Therefore, under complementation conditions, phage chromosomes can also recombine. It is important to realize, however, that complementation never requires the recombination of DNA molecules.

Complementation tests can also be carried out with bacteria, e.g., via the formation of partial diploids (merozygotes) by F-duction, or, with fungi, in a heterocaryon.

<u>Plan</u>. Five nonsense (<u>sus</u>) mutants of phage Kappa will be classified as allelic and non-allelic on the basis of their ability to complement each other. For this purpose, a lawn of non-suppressing (<u>su</u>) *Serratia* bacteria will be infected (spot test) with each of the possible pair combinations of the <u>sus</u> mutants. Lysis of the bacterial lawn means phage production and thus complementation. No lysis or the occurrence of only a few single plaques indicate that no complementation has taken place.

<u>Material</u>. First day: 1 ml stat culture in NB each of *S. marcescens* <u>su</u> W225 and <u>arg</u> su$^+$ W319 (cell titer approx. 1 × 10^9/ml). 3 ml of a suspension of each Kappa <u>sus</u> mutant, a, b, c, d and e, with plaque titers of 5 × 10^7/ml on indicator W319 and ≤5 × 10^3/ml on W225. 4 plates with NB agar. 4 NB soft agar tubes.

Procedure

<u>1st day</u>: Label NB agar plates (1-4) and mark 5 numbered sectors of equal size on the bottom of the plates. Layer plates 1-3 with soft agar and 0.2 ml each of indicator W225. Do the same with plate 4, but use indicator W319. Allow plates to stand open for about 10 min for drying. During this time, mix the <u>sus</u> mutants in 10 test tubes which had been previously labeled (ab; ac; ad; ae; etc.) according to the data sheet. The mixtures should contain 0.5 ml each of the first and the second <u>sus</u> mutant. Using a sterile inoculating loop, transfer one drop each of the mixed or the unmixed <u>sus</u> mutants, following the data sheet, on to the four NB plates. Allow the drops to dry entirely before incubating the plates at 30°C for 18 hrs.

Data sheet

Plate No.	Indicator bacteria	Plate sector	Complemen-tation partners	Complete lysis	Single plaques
1	W225 (su)	I	a + b		
		II	a + c		
		III	a + d		
		IV	a + e		
		V	b + c		
2	W225 (su)	I	b + d		
		II	b + e		
		III	c + d		
		IV	c + e		
		V	d + e		
3	W225 (su)	I	a solo		
		II	b solo	*Blank*	
		III	c solo		
		IV	d solo		
		V	e solo		
4	W319 (su$^+$)	I	a solo		
		II	b solo		*Blank*
		III	c solo		
		IV	d solo		
		V	e solo		

2nd day: Evaluate plates (Nos. 1 and 2) and control plates (Nos. 3 and 4):

1. Which of the sus mutants complement each other, i.e., which ones are non-allelic?

2. Which of the non-complementary and therefore "allelic" sus mutants are probably not isogenic and from which observation can this be concluded?

3. How can the behaviour of the phage mutant sus e be explained?

4. How could it be shown by a further test whether recombination has also occurred between phage chromosomes during the complementation of two non-allelic sus mutants?

Literature

GORINI, L.: Informational Suppression. Annual Review of Genetics 4, 107-134 (1970).

WINKLER, U., KOPP-SCHOLZ, U., HAUX, Ch.: Nonsense Mutants of Serratia Phage Kappa. Molec. Gen. Genetics 106, 239-253 (1970).

Time requirement: 1st day 1 hr, 2nd day 1 hr.

24. Constitutive and Induced Synthesis of β-Galactosidase

The operon theory of JACOB and MONOD explains cellular regulation processes acting at the level of transcription. The theory
was based on genetic and biochemical studies with *E. coli* bacteria.
In its modern and generalized form, the operon theory defines segments of DNA according to their function.

1. DNA segments which <u>code for the amino acid sequences</u> of
proteins. The phenotypic expression of these DNA segments requires transcription and translation. According to the function
of the protein, it is usual to distinguish between
tein, it is usual to distinguish between
a)"Structural" (S)-genes, which specify enzymes and structural
elements (e.g. the subunits of flagellae) and
b) "Regulatory" (R)-genes, which specify repressor molecules
that are not enzymatically active (for details see below).

2. DNA segments to which proteins of <u>specific amino acid se-</u>
<u>quences can bind</u> but which probably do not code for the synthe-
sis of amino acid sequences themselves.
a) Promotors (P-segments). DNA-dependent RNA polymerase recognizes promotors as specific binding and start sites for the transcription of one or more distal S- or R-genes.
b) Operators (O-segments). Active repressor molecules can specifically occupy operators. Thus, they prevent (negative control)
or promote (positive control) the transcription of the distal
S- or R-genes, beginning at the promotor.

3. DNA segments, which can be considered the <u>structural genes</u>
<u>for ribosomal RNA (rRNA) and transfer RNA (tRNA)</u>. These DNA
segments are transcribed but their RNA is not translated. The
transcription products participate in protein biosynthesis as
rRNA or tRNA.
While the DNA segments mentioned under (1) and (3) are generally
called genes, one still hesitates to classify the O and P segments as genes, especially since they are relatively short compared to "common" genes.

The best known "negatively controlled" operon is the "<u>lac operon</u>"
of *E. coli*.

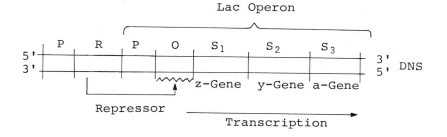

The z-gene codes for the catabolic enzyme β-galactosidase, which splits galactosides hydrolytically, e.g., lactose into glucose and galactose. The y-gene codes for the galactoside permease, which transports galactosides into the cells against a concentration gradient. The a-gene codes for the enzyme galactoside-trans-acetylase. If there is no suitable substrate (e.g., lactose) available, wild-type cells do not synthesize any of the three enzymes mentioned, because the repressor coded for by the R-gene (synonymous: i-gene) blocks the transcription of the genes z, y and a. If, however, substrate is offered, this substrate acts to "induce" (synonymous with derepress): When a substrate molecule complexes with a repressor molecule, the latter loses its high affinity for the operator (via an allosteric effect) and the transcription of the genes z, y and a begins.

In certain *E. coli* mutants, the above mentioned enzymes are formed at a high rate even if there is no substrate present; in "operator constitutive" (o^C) mutants, frequently short deletions, the operator is changed to such an extent that wild-type repressor no longer has an affinity for it. In "repressor constitutive" (R) mutants, the amino acid sequence of the repressor has changed to such a degree that the repressor no longer recognizes (and binds to) the operator.

The <u>lac</u> operon, which has been described above, is an example of the negative control of transcription. In addition to this, however, there are also "positively controlled" chromosomal segments in *E. coli*, e.g., the arabinose operon. Here, the product of the R-gene, activated by a substrate molecule, is necessary for transcription. Among the negatively controlled operons, one distinguishes between those whose enzymes have a catabolic (decomposing) function in the metabolism of the cells and those which have an anabolic (synthesizing) function:

Function of the operon	"Effector" of the regulation	Product of the R-gene Without effector	With effector	Behaviour of R mutants
Catabolic (e.g., lac)	Substrate ("Inducer") of an enzyme	active, i.e. repression of the enzyme	inactive, i.e. derepression synthesis	Continuous enzyme synthesis
Anabolic (e.g., trp)	End-product ("Corepressor") of a biosynthetic pathway	inactive, i.e. derepression of the enzyme	active, i.e. repression synthesis	

Outline of negatively controlled operons

<u>Plan</u>. A natural or a synthetic inducer of the <u>lac</u> operon will be added to an *E. coli* (wild type) culture, grown in inducer-free glycerol minimal medium. After incubation for different times, samples will be taken, treated with toluene and tested for their β-galactosidase activity. The increase of the β-galactosidase

activity with time is a measure of induction of the <u>lac</u> operon. As a control, a culture of a repressor-constitutive (R) mutant will be used which synthesizes the enzymes of the <u>lac</u> operon continuously, even in the absence of inducer.

<u>Material.</u> Approx. 50 ml each of a log culture in glycerol minimal (GM-)-medium of the *E. coli* strains Hfr H3000 (wild type) and Hfr H3300 (R mutant); cell titer, 3×10^8/ml each. 5 ml GM-medium. 50 ml P-buffer with 0.1 M β-mercaptoethanol. For inducer solutions, 2 ml of 0.1 M lactose and/or 1 ml of 0.01 thiomethyl galactopyranoside (MTG). As substrate for the enzyme assay, 4 ml of o-nitrophenyl-β-D-galactopyranoside solution (ONPG; 5 mg/ml). 2 500-ml Erlenmeyer flasks with lead rings to weigh down. 1 polystrene bucket with ice. 3 ml of toluene. 2 sheets of linear graph paper. 1 water bath at 37°C.

Procedure

1. <u>Enzyme induction and toluene treatment.</u> Pipette separately 50 ml each of the wild type and the mutant culture into the two 500-ml Erlenmeyer flasks and start incubation at 37°C in a water bath. Swirl the cultures continuously for aeration. Record all of the following measurements and time readings in data sheets I and II.

<u>Wild type.</u> At time t = 0 take a 5 ml sample from the corresponding flask and determine the cell density photometrically (O.D.$_{580}$) against sterile GM-medium. Immediately, add 0.5 ml of inducer solution (lactose or TMG), mix and continue incubation (induction). From the 5 ml sample which was taken to determine the cell density, put 2 ml into a small test tube containing 0.1 ml toluene ("tube 0") and discard the remainder. Shake "tube 0" vigorously, keep in 37°C water bath for exactly 1 hr and then chill in an ice bath. Repeat the procedure at t = 10 min, t = 20 min and t = 30 min ("tube 10, 20, or 30").

<u>Mutant (R).</u> At time t = 5 min take a 5 ml sample to measure the O.D.$_{580}$ and then treat it with toluene ("tube 5"). Some student groups should incubate the R culture with the inducer (lactose) and others without it. Again remove a sample at t = 35 min ("tube 35"), measure the O.D.$_{580}$ and treat with toluene. As soon as the tubes with toluene are in ice, begin with the β-galactosidase assay.

2. <u>β-galactosidase assay</u> (see data sheet I). The assay is based on the fact that β-galactosidase splits off the yellow-colored o-nitrophenol from the synthetic substrate ONPG, which is colorless in solution. The increase in the amount of o-nitrophenol with time can be followed photometrically at 420 nm.

<u>Assay mixture:</u> 4.0 ml P-buffer with β-mercaptoethanol

0.5 ml ONPG solution

0.5 ml toluene-treated cell suspension (without toluene!)

Pipette the solutions in this order into a cuvette, mix, place the cuvette immediately into the light path of a photometer and adjust the $O.D._{420}$ to zero. Then read the $O.D._{420}$ at intervals of 30 sec for 5 min and plot the values against the time (linear graph paper). Repeat this procedure with all 6 toluene tubes. "1 enzyme unit" (E.U.) is defined as the amount of enzyme which releases 1 μMol o-nitrophenol per min, corresponding to an increase of the $O.D._{420}$ of 0.0075 min^{-1} (for a cuvette with a 1 cm light path).

3. Evaluation (see data sheet II). For each enzyme assay, read the $\Delta_{O.D.}$ and the corresponding Δ_t in the region of linear increase of $O.D._{420}$ with time. List in the data sheets.

For each measured $O.D._{580}$ value of the experiment with the wild-type, calculate the ratio $F = O.D._{0\ min}/O.D._{x\ min}$ and also list in the data sheet. Calculate the same ratios from both $O.D._{580}$ values in the experiment with the mutant.

Calculate the enzyme units (E.U.) for each of the 6 tests as follows:

$$E.U. = \frac{\Delta_{O.D._{420}} \times 10}{\Delta_t \times 0.0075}$$

$\Delta_{O.D._{420}}$ = Increment of the $O.D._{420}$ in the time interval Δ_t (see data sheet).

10 = Dilution factor, as the initial cell suspension was diluted 1:10 at the beginning of the test.

0.0075 = Extinction of 1 μmole ONPG at λ = 420 nm and 1 cm light path.

List the calculated and uncorrected E.U. on data sheet II and multiply by the corresponding F-values: Now all of the E.U. refer to the cell density at t = 0. Plot the corrected E.U. graphically against the time of incubation in the presence of inducer.

Literature

BECKWITH, J.R., ZIPSER, D. (eds.): The Lactose Operon. Cold Spring Harbor: Cold Spring Harbor Laboratory 1972.

BOURGEOIS, S.: The Lac Repressor. In: Current Topics in Cellular Regulation 4, 39-75 (1971).

DICKSON, R.C., ABELSON, I., BARNES, W.M., REZNIKOFF, W.S.: Genetic Regulation: The Lac Control Region. Science 187, 27-35 (1975).

ENGELSBERG, E., WILCOX, G.: Regulation: Positive Control. Ann. Rev. Genetics 8, 219-242 (1974).

EPSTEIN, W., BECKWITH, J.R.: Regulation of Gene Expression. Ann. Rev. Biochem. 37, 411-436 (1968).

GILBERT, W., MÜLLER-HILL, B.: Isolation of the Lac Repressor. Proc. Natl. Acad. Sci. U.S. 56, 1891–1898 (1966).

ZUBAY, G., CHAMBERS, D.A.: Regulating the Lac Operon. In: Metabolic Pathways 5 (ed. H.J. VOGEL). London: Academic Press 1971.

Time requirement: 3.5 hrs.

Data sheet I (β-galactosidase assay)

Reading at t (min)	O.D.$_{420}$ of toluene-treated cell suspensions					
	0'	10'	20'	30'	5'	35'
0						
0.5						
1						
1.5						
2						
2.5						
3						
3.5						
4						
4.5						
5						
5.5						

Data sheet II

Removal of sample toluene addition		Cell density O.D.$_{580}$	End of toluene treatment[a] time	F	ΔO.D.$_{420}$	Δt min	Enzyme units uncorrected (E.U.)	corrected (E.U. $\times$ F)
t min	time							
Wild type								
0								
10								
20								
30								
Mutant								
5 $\triangleq$ 0								
35 $\triangleq$ 30								

[a] The toluene-treated cell suspensions have to be kept in an ice bath.

$$F = \frac{\text{O.D.}_{580} \text{ at time } 0}{\text{O.D.}_{580} \text{ after } t \text{ minutes of induction}}$$

25. The Morphopoesis of T4 Phages

Genetic experiments have shown that approx. 46 of the 70 known genes of phage T4 participate in its morphopoesis. In contrast, only 15-20 different structural proteins appear in mature phage particles, as demonstrated by biochemical methods and by electronmicroscopy. This discrepancy can be explained by the assumption, that some steps in the assembly of the structural components require proteins with catalytic functions. Some of these proteins are probably "scaffolding devices" which may function as morphopoetic cores. Others may activate individual subunits, e.g. by specific cleavage, which facilitates their subsequent spontaneous assembly.

Using nonsense mutants (cf. Expt. 23) of phage T4, some steps of phage morphopoesis can be studied: if nonsuppressing host bacteria (su) are infected with phages which have a sus mutation in one of the morphopoetical genes, instead of whole, infectious phage progeny, only pieces of phages, such as heads, tails, tail fibers or other structures, accumulate in the cells. If cell free extracts of su bacteria which were infected with different sus mutants are mixed, under suitable conditions the accumulated phage subunits of both cell extracts complement mutually, and infectious phage particles are formed *in vitro*. This proves that the phage subunits, which have been observed in the electron microscope, are not a result of a defective synthesis, but are intermediate products of phage formation. The *in vitro* complementation of phage mutants gave insight into the pathway of assembly of phage T4: at first the subunits of the head, tail and tail fibers are synthesized independently. Then the head and tail subunits react with one another and, only after this, the tail fibers are able to attach. This sequential self-assembly of the phage particles indicates that some gene products must react with certain intermediary structures, to condition them to be a substrate for the next step.

The morphopoesis of phage T4 is at present the bestunderstood model of sequential self-assembly of supramolecular structures starting from a large number of gene products.

Plan. Two cultures of *E. coli* BA (su) will each be infected with one of the two T4 sus mutants am N120 (head donor) and am B17 (tail donor). The cells of both cultures will be concentrated by centrifugation and will then be disintegrated by quick freezing and thawing. The cell extracts thus obtained will be mixed and the increase of infectious phages, caused by morphopoesis, will be determined on su^+ bacteria as indicator.

Material (Part 1). 60 ml of a stat culture in HERSHEY broth (37°C, aerated) of *E. coli* BA (cell titer, approx. 3×10^9/ml). 0.5 ml each of the phage suspensions T4 am N120 and T4 am B17 (titer, 1×10^{12}/ml). Two 500-ml Erlenmeyer flasks containing 175 ml HERSHEY broth each, with aeration devices. 5 ml of a tryptophan solution, 100 mg/ml. 2 1000-ml Erlenmeyer flasks. 1 plastic bucket with ice. 2 250-ml centrifuge bottles. 2 plastic tubes. 2.5 ml ice-cold phosphate buffer with 10 µg/ml DNAse (Calbiochem, B grade). 2 PASTEUR pipettes. Sterile absorbent paper. For several groups together: One SORVALL centrifuge, model RC2-B, with a GSA rotor. One 37°C and one 30°C water bath. One methanol bath, cooled to -70°C. 10 ml each of a stat culture of *E. coli* BA and *E. coli* CR63, both with a cell titer of approx. 3×10^9/ml. Part 2: 32 plates with EH agar. 32 EH soft agar tubes. 300 ml P-buffer. For several groups together: 1 water bath at 30°C.

Procedure

Part 1: Preparation of cell extracts. Inoculate 2 Erlenmeyer flasks, each containing 175 ml HERSHEY broth, with 25 ml each of a stat culture of *E. coli* BA and aerate at 37°C until a cell titer of 4×10^8/ml is reached (O.D.$_{580}$ = 0.4). Transfer both cultures to a 30°C water bath and immediately add 2 ml of the tryptophan solution and 0.3 ml T4 am N120 or T4 am B17; this corresponds to an average multiplicity of infection of about 4. Aerate both cultures vigorously at 30°C for 30 min and then quickly chill each culture separately in a pre-cooled 1000 ml Erlenmeyer flask. Centrifuge the cultures immediately in 250-ml centrifuge bottles at 5,500 rpm (5,000 × g) in the GSA rotor for 8 min. Discard the supernatants and dry the contents of the centrifuge bottles well by use of PASTEUR pipettes and sterile absorbent paper, in order to remove as much of the non-adsorbed phages as possible (background!). Take up each of the viscous sediments in 1 ml of phosphate buffer containing DNAse; the cultures have now been concentrated 200 fold. Put the bacterial suspensions, infected with T4 am N120 or T4 am B17, into two plastic tubes and quickly freeze in a menthanol bath at -70°C; in order to desintegrate the cells first thaw at 30°C, and freeze once more. By this treatment, 99% the suspension of the cells are destroyed. Check microscopically (magn. 10 × 40).

Part 2: Titers of the cell extracts. Label plates according to data sheet I and prepare the dilution rows with P-buffer. Thaw both cell extracts at 30°C and mix 0.5 ml of one with 0.5 ml of the other suspension. Immediately take 0.1 ml samples from all 3 extracts (N120, B17 and mixture), dilute according to data sheet I and plate on su and su$^+$ indicator with the soft agar layer technique. While this is being done, incubate the 3 cell extracts at 30°C for 200 min. Then again dilute aliquots of 0.1 ml according to the data sheet and plate. Incubate all agar plates at 30°C overnight.

Data sheet I

Be careful not to mix the indicators!

Extract of BA and	Dilution	Plate-No. and indicator BA($\underline{su}$) CR63($\underline{su}^+$)		Plaque titer	Label
Time: 0 min					
am N120	10^{-2}	1	–		A_0
	10^{-2}	2	–		
	10^{-5}	–	3		
	10^{-5}	–	4		B_0
	10^{-6}	–	5		
am B17	10^{-2}	6	–		C_0
	10^{-2}	7	–		
	10^{-5}	–	8		
	10^{-5}	–	9		D_0
	10^{-6}	–	10		
am N120 + am B17	10^{-2}	11	–		E_0
	10^{-2}	12	–		
	10^{-3}	13	–		
	10^{-6}	–	14		
	10^{-6}	–	15		F_0
	10^{-7}	–	16		
Time: 200 min					
am N120	10^{-2}	17	–		A_{200}
	10^{-2}	18	–		
	10^{-5}	–	19		
	10^{-5}	–	20		B_{200}
	10^{-6}	–	21		
am B17	10^{-2}	22	–		C_{200}
	10^{-2}	23	–		
	10^{-5}	–	24		
	10^{-5}	–	25		D_{200}
	10^{-6}	–	26		
am N120 + am B17	10^{-2}	27	–		E_{200}
	10^{-2}	28	–		
	10^{-3}	29	–		
	10^{-6}	–	30		
	10^{-6}	–	31		F_{200}
	10^{-7}	–	32		

Evaluation

Count the plaques on plate Nos. 1-32 and calculate the titers. The titers A, C and E represent the back mutants ($\underline{sus} \longrightarrow \underline{sus}^+$) in the corresponding extracts and the titers B, D and F the sum of all the infectious T4 phages.

According to data sheet II, calculate ratios from the different titers and discuss.

Data sheet II (Evaluation)

	Ratio	Frequency of $\underline{sus}^+$ back mutants at 0 and 200 min
$\underline{am}$ N120	A_0/B_0 A_{200}/B_{200}	
$\underline{am}$ B17	C_0/D_0 C_{200}/D_{200}	
$\underline{am}$ N120 + $\underline{am}$ B17	E_0/F_0 E_{200}/F_{200}	
$\underline{am}$ N120 + $\underline{am}$ B17	Ratio	Frequency of infectious phages, formed by *in vitro* morphopoesis
Indicator CR63	$2 \times F_0/(B_0 + D_0)$ $2 \times F_{200}/(B_{200} + D_{200})$	
Indicator BA	$2 \times E_0/(A_0 + C_0)$ $2 \times E_{200}/(A_{200} + C_{200})$	

Literature

EDGAR, R.S., WOOD, W.B.: Morphogenesis of Bacteriophage T4 in Extracts of Mutant-Infected Cells. Genetics 55, 498-505 (1966).

KIKUCHI, Y., KING, J.: Genetic Control of Bacteriophage T4 Baseplate Morphogenesis. J. Molec. Biol. 99, 645-672 (1975).

KUSCHNER, D.J.: Self-Assembly of Biological Structures. Bacteriological Rev. 33, 302-345 (1969).

WEIGLE, J.: Studies on Head and Tail Union in Bacteriophage Lambda. J. Molec. Biol. 33, 483-489 (1968).

Time requirement: 1st day (Parts 1 and 2) 7.5 hrs, 2nd day 2 hrs.

Problems

1. How many different amino acids could be coded for by an mRNA which is statistically composed of the 4 usual nucleobases, if the genetic code
a) were a non-degenerate doublet code?
b) were a non-degenerate triplet code?

2. Polyribonucleotides can be synthesized *in vitro* and can then be used as mRNA. If the initial mixture for the synthesis of a poly-ribonulceotide consists of 70% adenosine (A)- and 30% cytidine-(C)-5'diphosphate and if polynucleotide phosphorylase adds A and C at random
a) which triplets can be formed?
b) what is the chance of their formation?
c) which amino acids are coded by the synthetic polynucleotides?

(Note to Question b: The probability that $\geqslant 2$ independent events occur at the same time equals the product of the probabilities for each individual event.)

3. The alanine and arginine content of the total protein is sig-nificantly higher in certain bacterial types than in others. What conclusions can be drawn concerning the average GC content of the DNA of these bacteria?

4. Under favorable growth conditions, the generation time of *E. coli* is 30 min; the molecular weight of the *E. coli* chromosome is about 3×10^9 daltons.
a) How many nucleotide pairs are inserted per second during replication, assuming that replication occurs continuously and at a constant rate?
b) Why does the transcription of *E. coli* DNA proceed at a much slower rate? (About 30 nucleotides/sec.)

5. Let us assume that bacteria are diploid for the structural genes a and b, the regulatory gene R and the operator o. The proteins A and B, synthesized in a^+ and b^+ strains, are cata-bolic enzymes (such as β-galactosidase). Their synthesis is inducible by substrate ("inducer"). The repressor, coded by R^+, blocks only o^+, and is inactivated by the substrate mole-cule. Under which conditions are the proteins A and B formed?

R o a b / R o a b	Synthesis of protein			
	without inducer		with inducer	
	protein A	protein B	protein A	protein B
− + − − / + − + +				
− + + + / + − − −				
+ + − + / + + − +				
+ − − + / + + + −				

Note: o⁻ means repressor does not bind to this DNA (usually symbolized as o^C).

6. Many antibiotics specifically inhibit the metabolism of the bacterial cell. What effects do the following antibiotics have on bacterial cells, and, if known, where do they attack?

a) Actinomycin D

b) Chloramphenicol

c) Mitomycin C

d) Penicillin G

e) Puromycin

f) Rifampicin

7. In bacteria a single point mutation simultaneously can result in several changes in the phenotype, e.g.,
a) polyauxotrophy for tryptophane, tyrosine and phenylalanine;
b) loss of the ability to utilize several different sugars as carbon and energy source. How can these pleiotropic mutations be explained?

III. Appendix

A. Nutrient Media and Solutions

Autoclave all nutrient media at 121°C for 15 min. Media which contain 1-2% agar are poured into plastic petri dishes in 20 ml portions (agar plates). Soft agar, immediately after autoclaving, is distributed in 3 ml portions into small test tubes and kept molten in a 47°C water bath. If a medium is to be supplemented with a material whose heat stability is doubtful (e.g., antibiotics), it is recommended that concentrated stock solutions of the material be prepared in sterile water or buffer. Add suitable quantities to the media after autoclaving.

All nutrient media and all solutions used for the dilutions listed under "Material" have to be sterile. All the enzyme, antibiotic or mutagen solutions used, whose activity might decrease at room temperature, are to be freshly prepared in buffer or sterile distilled water before the start of the experiment. Percentages of media and solutions are always given in percent of weight.

Broth (NB)

8 g nutrient broth
4 g NaCl
to 1,000 ml with distilled water

Tryptone yeast broth (TBY)

10 g tryptone
 5 g yeast extract
 5 g NaCl
to 1,000 ml with distilled water

NB Agar

15 g agar
to 1,000 ml with NB

TBY Agar

15 g agar
to 1,000 ml with TBY

NB Soft agar

5 g agar
to 1,000 ml with NB

TBY Soft agar

5 g agar
to 1,000 ml with TBY

HERSHEY broth (HB)

8 g nutrient broth
5 g peptone
5 g NaCl
1 g glucose
to 1,000 ml with distilled water

Enriched HERSHEY agar (EHA)

10 g agar
13 g tryptone
 8 g NaCl
 2 g Na-Citrate × 2H$_2$O
1.3 g glucose
to 1,000 ml with distilled water

HB Agar

15 g agar
to 1,000 ml with HB

EH Soft agar

EHA medium, but
6.5 g instead of 10 g agar
3 g instead of 1.3 g glucose

Peptone glycerol (PG) agar

 5 g peptone
10 ml glycerol
20 g agar
to 1,000 ml with distilled water

Minimal medium (M9)

Solution 1: 20 g glucose
            to 500 ml with distilled water

Solution 2: 2.5 g $MgSO_4$ × $7H_2O$
            to 100 ml with distilled water

Solution 3: 0.2 g $CaCl_2$ × $2H_2O$
            to 100 ml with distilled water

Solution 4: 750 ml distilled water

Solution 5: 35 g Na_2HPO_4 × $2H_2O$
            15 g KH_2PO_4
             2.5 g NaCl
             5 g NH_4Cl
            to 500 ml with distilled water

After autoclaving, mix

750 ml solution 4
100 ml solution 1
100 ml solution 5
 10 ml solution 2
 10 ml solution 3

M9 Agar

Like M9, but Solution 4 contains 15 g agar

Supplemented M9 agar (M9s)

970 ml M9 Agar + 5 ml NB

Glycerol minimal (GM) medium

Like M9, but Solution 1 contains 0.3% casamino acids and 25 g glycerol instead of 20 g glucose. Add thiamine hydrochloride to a final concentration of 0.5 µg/ml.

Eosin methylene blue agar (EMB)

68.6 g EMB mix (see below)
to 900 ml with distilled water
After 15 min autoclaving, add
100 ml sterile aqueous solution containing
20% lactose or
15% maltose or
10% galactose or other sugars

EMB Mix

 5   g K_2HPO_4
12.5   g NaCl
 2.5   g yeast extract
25   g tryptone
40   g agar
 0.125 g methylene blue B
 0.75 g eosin yellow

RYAN's supplemented minimal agar

1,000 ml Solution A ←	3 g K_2HPO_4
10 ml solution of trace elements	1 g KH_2PO_4
(see below)	5 g NH_4Cl
10 ml 50% glucose	1 g NH_4NO_3
10 ml 15% asparagine	2 g Na_2SO_4
1 ml 1% $CaCl_2 \times 2H_2O$	20 g agar
1 ml 10% $MgSO_4 \times 7H_2O$	to 1,000 ml with distilled
1 ml 0.1 % L-histidine-HCl	water

Trace element solution

1,000 mg Fe (III) citrate	1.0 mg $Na_2MoO_4 \times 2H_2O$
1,000 mg $CaCl_2 \times 2H_2O$	5.0 mg $CoCl_2$
10 mg $MnCl_2 \times 4H_2O$	0.5 mg $SnCl_2 \times 2H_2O$
5 mg $ZnCl_2$	0.5 mg $BaCl_2$
0.5 mg LiCl	1.0 mg $AlCl_3$
2.5 mg KBr	10.0 mg H_3BO_3
2.5 mg KI	20.0 mg ethylenediaminetetra-
0.005 mg $CuSO_4$	acetate (EDTA)
	to 1,000 ml with distilled water

Some of the following buffers are concentrated stock solutions, from which the desired concentration can be prepared by dilution.

Tris-HCl buffer (1M)

121 g Tris (hydroxymethyl)-aminomethane
800 ml distilled water

Adjust to desired pH with 1 N HCl and fill to 1,000 ml with distilled water.

20-fold standard saline citrate buffer (20 × SSC)

176 g NaCl (corresponds to 3M)
 88 g sodium citrate (corresponds to 0.3 M)
to 1,000 ml with distilled water

Phosphate buffer, neutral (P buffer)

Solution 1: 7 g $Na_2HPO_4 \times 2H_2O$ (corresponds to 0.04M)
            3 g KH_2PO_4 (corresponds to 0.02M)
            4 g NaCl
            to 1,000 ml with distilled water

Solution 2: 2.5 g $MgSO_4 \times 7H_2O$
            to 10 ml with distilled water

Autoclave for 15 min; after cooling, 1,000 ml solution 1 + 2 ml of solution 2

Potassium acetate buffer (1M)

 98 g potassium acetate
800 ml distilled water

Adjust to pH 5.2 with concentrated acetic acid and fill up to 1,000 ml with distilled water

Toluene scintillator

5 g   2,5-diphenyloxazole (PPO);
0.22 g 2,2'-p-phenylene-bis-(4-methyl-5-phenyl oxazole)(POPOP);

Fill up to 1,000 ml with toluene

Dioxane scintillator

 60 g naphthaline
  4 g PPO
 0.2 g POPOP
100 ml methanol
 20 ml ethylene glycol

Fill up to 1,000 ml with dioxane

B. Strains of Bacteria and Phages

Bacterial strains	Synonym	Needed for Expt.No.[a]
Escherichia coli		
BA <u>su</u>	W590	1, 2, 3, (7), 8, 10, 25
K12s	W1033	4, (18), 20
15 <u>his</u>	W484	12
GY767 Hfr <u>str</u>^{s}	W1034	16
K12 (λ)λ^r	W993	20
W3100 (<u>col</u>D-20)	W789	20
W3110 <u>str</u>^{r}<u>azi</u>^{r}<u>nal</u>^{r}	W779	20
C600 (λc$_I$857)<u>thr</u> <u>leu</u> <u>thi</u> <u>tonA</u> <u>lac</u>	W1032	(21)
Hfr H3000 Wild type	W574	19, 24
Hfr H3300 <u>thi</u> <u>i</u> (synonymous R⁻)	W575	24
CR63 <u>su</u>$^+$ amber	W473	10, 25
K12 <u>str</u>^{s}<u>thi</u> (<u>lac</u> <u>pro</u>) Deletion F'<u>lac</u>$^+$<u>pro</u>$^+$	W1023	19
AB1157 F⁻λ^s <u>thr</u> <u>leu</u> <u>arg</u> <u>pro</u> <u>his</u> <u>thi</u> <u>str</u>^{r} <u>ara</u> <u>xyl</u> <u>mtl</u> <u>lac</u> <u>gal</u>	W1022	16, 18, 19, 21
AB1157, Gene markers as above, but λ^r	W1031	21
Serratia marcescens		
HY wild type <u>su</u>	W225	11, 13, 15, 17 22, 23
HY <u>leu</u>	W315	11
HY <u>pig</u> <u>prt</u>	W226	11
HY <u>hcr</u>$_{42}$	W227	13
HY <u>arg</u> <u>su</u>$^+$	W319	17, 23
HY <u>thi</u>	W366	15, 22

---

[a] see p. 172.

Bacterial strains	Synonym	Needed for Expt.No.[a]
Serratia marcescens		
HY thy 1	W626	22
HY his 26	W861	22
HY lys 69	W982	22
HY pig	W592; P18	22
HY pig 9-3-3	W623	22
Nima OF	W622	22

Phage strains	Synonym	Needed for Expt.No.[a]
E. coli phages		
T4 wild type	PW246	6, 8, (9)
T4rII73	PW108	1, 2
T4o	PW105	1, 2, 3
T4 am N120 (Gene 27)	PW247	25
T4 am B17 (Gene 23)	PW244	25
T4 am N91 (Gene 37)	PW245	10
T5 wild type	PW135	6, (9)
λwild type	PW138	4
λcb$_2$b$_5$	PW139	4
λc$_I$857, obtained from appropriate lysogenic strains, e.g. W1032		(21)
Plkc	PW252	18
Serratia phages		
κwild type	PW196	13
κsus L90	PW90;a;x	17, 23
κsus N95	PW95;b;y	17, 23
κsus O86cl	PW155;c;z	17, 23
κsusO86 susL90	PW195;e;xz	17, 23
κsus O191	PW191;d	23

[a] Numbers in parenthesis indicate that the respective strain is not required directly in the experiment but rather is needed for the preparation of material for the experiment.

Explanation of the Symbols Used for the Characterization of the Genotypes

Auxotroph for

arg arginine

his histidine

leu leucine

lys lysine

pro proline

thi thiamine

thr threonine

thy thymine

Non-fermenting

ara arabinose

gal galactose

lac lactose

mtl mannitol

xyl xylose

(i   lac repressor)

Resistant to

azi^r sodium azide

nal^r nalidixic acid

str^r streptomycin

ton phage T1

F^- lacking fertility factor

Hfr with fertility factor, integrated (high frequency of recombination)

F' with sex-ducing fertility factor (F prime)

Others: defective in

hcr host cell reactivation

pig prodigiosin (pigment)- synthesis

prt protease synthesis

su   suppression of nonsense triplets

Gene markers of phages

am   amber (nonsense) triplet

b2,b5 deletions

c   clear plaque

o   resistance to osmotic shock

r   rapid lysis

sus   nonsense triplet (suppressor sensitive)

Behaviour towards plasmids/temperate phages

col D colicinogenic for colicin D

(λ)   lysogenic for λ

λ^r adsorption resistant to λ

λ^s lyso-sensitive for λ

C. Sources of Supplies for Experiments

Media

Difco Laboratories
920 Henry Street
Detroit, MI 48201, USA

Oxoid Ltd.
Southwark Bridge Rd.
London, E.C.4, England

Biochemical Reagents and Enzymes

Aldrich Chem. Co., Inc.
2369 North 29th St.
Milwaukee 10, WI, USA

Boehringer-Mannheim GmbH
Biochemica
Postfach 51
6800 Mannheim 31, FRG

Calbiochem
P.O. Box 12087
San Diego, CA 92112, USA

Fluka AG
Chemische Fabrik
9470 Buchs, Switzerland

Hoffmann-La Roche
Nutley, NJ 07110, USA

E. Merck AG.
Frankfurter Str. 250
6100 Darmstadt, FRG

Merck Chemical Division
Merck and Co., Inc.
Rahway, NJ 07065, USA

Miles Laboratories, Inc.
Research Products Department
1127 Myrtle Street
Elkhart, IN 46514, USA

Pharmacia Fine Chemicals
Uppsala, Sweden

Serva Feinbiochemica GmbH & Co.
Postfach 1505
6900 Heidelberg 1, FRG

Sigma Chem. Comp.
3500 De Kalb St.
St. Louis, MO 63118, USA

Union Carbide Corp.
6733 West 65th St.
Chicago, IL 60638, USA

Worthington Biochemical Co.
Freehold, NJ 07728, USA

Radiochemicals

New England Nuclear
575 Albany St.
Boston, MA 02118, USA

The Radiochemical Center
Amersham, Buckinghamshire, England

Equipment

Centrifuges	Ivan Sorvall, Inc. Pearl Street Norwalk, CT 06856, USA
Ultra- centrifuges	Beckman Instruments, Inc. Scientific Instr. Division 2500 Harbor Boulevard Fullerton, CA 92634, USA
Spectrophoto- meters	Carl Zeiss 7082 Oberkochen, FRG Bausch & Lomb Scientific Instr. Division 635 St. Paul Street Rochester, NY 14602, USA
Scintillation counters	Packard Instruments Comp., Inc. 220 Warrenville Road Downers Grove, IL 60515, USA

Glasware and Miscellaneous

Disposable µl pipettes	Corning Glass Works Corning, NY 14830, USA
Filters	Schleicher and Schuell, Inc. Keene, NH 03431, USA

Bacterial & Phage Strains used in this textbook

Ruhr-Universität
Lehrstuhl Biologie der Mikroorganismen
Postfach 2148
4630 Bochum, FRG
(Supplied at a minimum charge)

D. The Recording of Scientific Experiments

The important points, those which should be noted for an accu-
rate record of an experiment, have been put together in the following pages. As a rule, an experienced researcher would rarely
make such extensive records, but a beginner should keep more extensive notes so that later, when a "telegram style" is used, he
will have learned not to omit important data.

I. Planning, Preparation and Execution of the Experiment

1. Generalities

Write down the formulation of the problem and the intended plan
of an experiment before beginning. A time interval between this
planning and the actual experiment aids in the detection of
omissions or mistakes. During the experiment, immediately record observations and measurements. Do not make rough drafts,
because mistakes can occur during the rewriting. Evaluate the
data sheets immediately after the experiment. Keep agar plates,
solutions, etc, until after the evaluation, as they might be
needed for rechecking questionable results.

2. Contents of a Record

Date. The actual day of a recording.
Number of pages. The successive numbering of the pages eases
the recording. A bound record book is preferable to a collection
of separate sheets.
Title. Either use the experimental problem or the method and
the test object, e.g., "Buoyant Density of Phage X" or "CsCl
Gradient Centrifugation of Phage Y".
Formulation of the question and principle of the experiment,
when not clear from the title.
Reference. Either quote the scientific literature which refers
to the problem or the method, or point out your own preliminary
data (data book, pages and date).
Test objects. Name and origin of the organisms or viruses, e.g.,
RNA phage f2 (ZINDER), should be recorded.
Reagents, nutrient media, enzymes, instruments, etc.
By frequent use, an abbreviation is sufficient, e.g., "NB" for
DIFCO nutrient broth. When referring to special equipment,
specify manufacturer, distributor and charge number.
Course of experiment. The preparations, e.g., culturing of bacteria or dialysis conditions of DNA, and the experimental plan,
e.g., schedule for removal of samples and numbering of tubes,
is to be listed. If possible, the course of the experiment should
be recorded in form of a table and room should be allowed for
recording measurements, calculations, etc., which might follow.

II. Evaluation

1. Qualitative
a) Classify observations and define the "classes" well, even when the lack of clearly defined boundaries makes the selection arbitrary.
b) If a result is not clear, record the subjective estimation nevertheless, but be careful in the formulation (e.g., "It appears that..."). Routinely, avoid evaluating an experiment schematically, because otherwise the unexpected is easily overlooked.
c) "Check" observations with alternative measures, e.g., do the data correspond to the theoretical expectations, can they be explained by random fluctuation, can they be a result of contamination (bacterial, radioactive, etc.)? This point must be taken into consideration for quantitative evaluation, as well.

2. Quantitative
a) Record measurements or counts and list the results in tables. Clearly mark values which had to be estimated (approx. ...).
b) Calculate results and pay attention to
- Rechecking, i.e., note all calculation steps.
- Comparability of the values within a test or with those of an earlier test, i.e., when needed, change absolute numbers into per cent values, etc.
- Statistical verification of a statement, i.e., note standard deviations and include special statistical procedures (t-test, Chi^2 test, regression analysis and others).

3. Discussion of the Results
a) Does the experiment answer the basic problem or part of it? If not, was the problem too complicated or was the design of the experiment inadequate?
b) Attempt to explain the results. When a hypothesis already exists before the beginning of the experiment, check whether the results confirm the hypothesis or whether it must be modified, etc. Unusual interpretations are better than none at all, but simple hypotheses are preferable to more complicated ones.
c) Immediately consider plans for new experiments, based on the one just completed, while the results and experimental conditions are still in mind. If the data of one experiment appear to be important, immediately repeat the experiment.
d) Quote publications dealing with related experimental questions, techniques or results applying to your experiment (journal, year, volume, pages, publishers, abbreviated title).

Literature

SQUIRES, G.L.: Practical Physics. New York: McGraw Hill 1968.

IV. Experimental Results and Answers to the Problems

A. Experimental Results

Experiment 1 (T4 Production)

Two identical *E. coli* cultures were incubated; one was infected with T4o and the other with T4rII73. The multiplicity of infection was approx.

$$\frac{1 \text{ ml} \times (2 \times 10^{10}/\text{ml})}{1000 \text{ ml} \times (4 \times 10^{8}/\text{ml})} = 0.05.$$

While the T4rII73 infected culture lysed, the cell density of the T4o infected culture did not decrease, probably due to "lysis inhibition". After chloroform addition and centrifugation the phage titer in the crude lysate was:

$1.2 \times 10^{11}/\text{ml}$ T4o   and

$0.3 \times 10^{11}/\text{ml}$ T4rII73.

The difference is based on the fact that lysis inhibition increases the average phage yield per infected cell (burst size).

Growth constant k (calculated according to cell titer between the first and second hour of the experiment)

$$k = \frac{2.3(8.85 - 8.40)}{60} = \underline{\underline{1.74 \times 10^{-2} \text{ min}^{-1}}}.$$

Number of generations

$$g = \frac{1.74 \times 10^{-2} \times 60}{0.69} = \underline{\underline{1.5}}.$$

Generation time

$$T = \frac{60}{1.5} = \underline{\underline{40 \text{ min}}}.$$

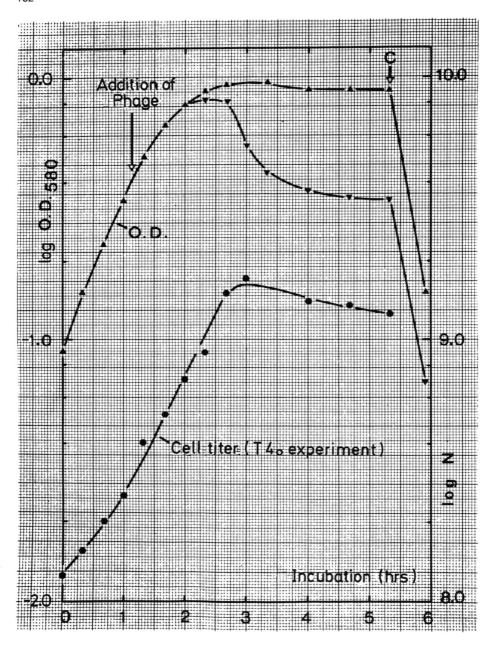

Growth curves of *E. coli* BA before and after infection with T4o
(▲) or T4rII73 (▼), respectively. *C* Addition of chloroform
followed by centrifugation

Experiment 2 (Concentration of T4 by the Dextran-PEG-Technique)

- Most of the phages are removed from the upper phase with 0.9 ml PEG solution. The partition coefficient would be

$$\frac{C_u}{C_l} = \frac{2.6 \times 10^6}{4.5 \times 10^8} = \underline{5.8 \times 10^{-3}}.$$

- The final molar concentration of NaCl in the test tube with 0.9 ml PEG solution is 0.25 M.

- In this test, 0.9 ml PEG solution is sufficient for the separation. With 1.2 ml PEG solution, separation would be complete. Assuming that 1 g of 30% PEG approx. equals a volume of 1 ml, then there are

  0.36 g PEG/1.2 ml solution   or

  0.36 g PEG/3.6 ml T4 lysate

  $$\frac{10,000 \text{ ml} \times 0.36 \text{ g}}{3.6 \text{ ml}} = \underline{1,000 \text{ g PEG/10 L.}}$$

Experiment 3 (Concentration of T4 by Centrifugation)

1. Factor of concentration and yield:

   $A/B = (9.7 \times 10^{12})/(1.2 \times 10^{11}) = 81$

   $C/D = \quad 640 \text{ ml}/3.7 \text{ ml} \quad = 173$

   $81/173 = 0.47$, i.e., the yield amounts to about 47%.

2. Purity of the T4o suspension

Ratio	Suspension of dialyzed phages	DNA solution
260/280	0.264/0.181 = 1.46	1.89
260/235	0.264/0.134 = 1.97	2.22

DNA in aqueous solution has a maximum of adsorption near 260 nm and a minimum at about 235 nm whereas the corresponding values for a solution of protein are about 280 nm and 245 nm, respectively. Since phages consist of DNA plus protein the O.D. ratios of a suspension of purified phage are lower than those of a DNA solution. The O.D. ratio 260 nm/280 nm is about 2 for DNA and about 0.5 for protein. The $O.D._{260}$ for 4.25×10^{10} phages/ml was 0.264. Thus, a suspension of 1×10^{12} phages/ml would correspond to an $O.D._{260}$ of 6.25.

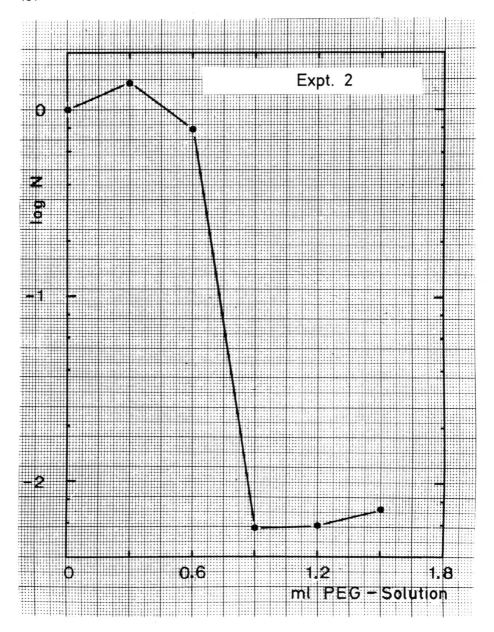

Polyethylene glycol (PEG) - dextran - two-phase technique.
Relative titer ($N_{corrected}$) of phage T4 in the supernatant as a
function of the amount of PEG solution

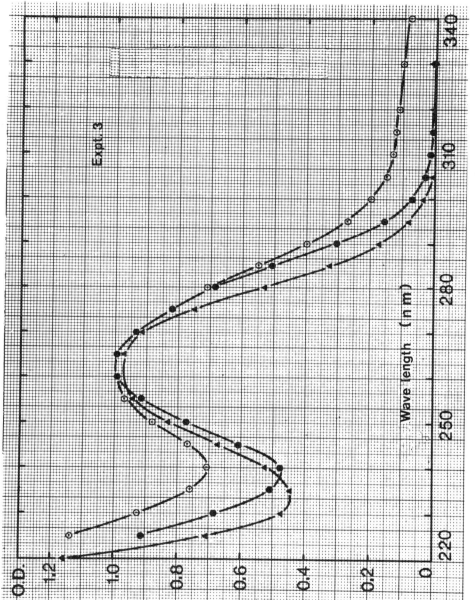

Normalized absorption curves of a suspension of purified T4 phages and a T4 DNA solution (▲) in the range of λ = 220-340 nm. The phage curves are uncorrected (o) or corrected (●) for light scattering, respectively

Experiment 4 (CsCl Gradient Centrifugation)

Phage	Refractive index η	Density ρ
Wild type	1.3812	1.502
cb_2b_5	1.3791	1.481

$\Delta\rho$ 0.021

Hence, $\alpha = \dfrac{2 \times 0.021}{0.21 - 0.021} = \underline{\underline{0.222}}$.

The difference between the DNA content of λcb_2b_5 and that of the wild type phage amounts to 22.2% (published value approx. 18%).

Directions for the instructor of the course

a) After a centrifugation time of about 24 hrs the equilibrium state of the CsCl gradient is not yet complete; however, the conditions suffice for the purposes of this course.

b) The volume of a drop and thus the number of drops per gradient depends on the inner diameter of the hypodermic needle; therefore, the number of drops which are to be collected per fraction should be ascertained in a preliminary test, in order to obtain a total of about 30 fractions.

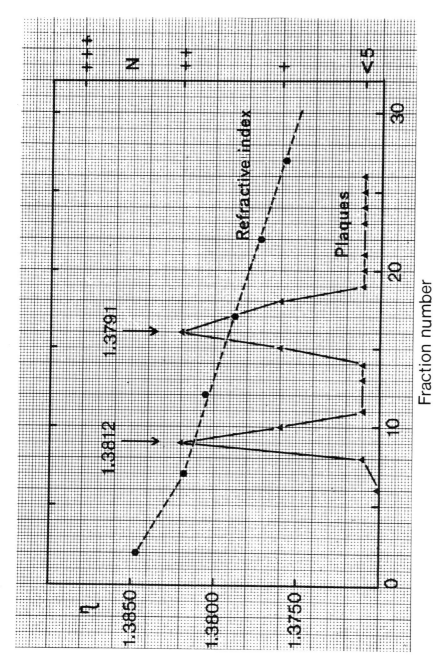

CsCl density gradient centrifugation of phage λ wild-type and λcb₂b₅. Relative phage titer N (▲) and the refractive index η_{250} (●) as a function of the fraction number

Experiment 5 (Sucrose Gradient Centrifugation)

Type of evaluation	Reference molecule	E_1/E_2	s-value $_{LDH}$
graphic	cyt. + βgal	–	7.2 S
calculated	cyt.	10/3 = 3.33	5.66 S
calculated	βgal $_I$	10/21 = 0.48	7.6 S
calculated	βgal $_{II}$	10/31 = 0.32	7.26 S

The graphic evaluation uses the given s-values of the peaks of the respective proteins, cytochrome c, β-galactosidase and LDH which are easily determined. 7.2 S, the s-value, which was obtained graphically for LDH is close to that which is quoted in the literature, s = 7.055.

The mathematical evaluation of the cytochrome c peak is not very exact because this peak is too close to the meniscus (this holds for the graphic evaluation, too). For more exact determinations of the s-value, we recommend the use of proteins which sediment faster than cytochrome c or, if these are not available, longer centrifugation times. Reduction of the volume of the fractions might also help. For the purpose of this course, however, cytochrome c, because of its red color, allows the instructor to judge immediately whether the gradient was properly formed and whether the sample was applied carefully.

Molecular weights:

cytochrome c	12,500
LDH	131,000
β-Galactosidase (from *E. coli*)	521,000

Continuation of Experiment 5

Fraction No.	O.D.$_{420}$ (Cytochrome c)	ΔO.D.$_{340}$ (LDH)	O.D.$_{420}$ (β-Gal)
1	–	–	0.035
2	–	–	0.070
3	–	–	0.125
4	–	–	0.128
5	–	–	0.113
6	–	–	0.083
7	–	–	0.075
8	–	–	0.058
9	–	–	0.045
10	–	–	0.060
11	–	–	0.080
12	–	–	0.213
13	–	–	0.335
14	–	–	0.390
15	–	–	0.325
16	–	–	0.168
17	–	–	0.085
18	–	0.002	–
19	–	0.002	–
20	–	0.010	–
21	–	0.008	–
22	–	0.015	–
23	–	0.065	–
24	–	0.128	–
25	–	0.203	–
26	–	0.145	–
27	–	0.085	–
28	–	0.035	–
29	0.135	–	–
30	0.228	–	–
31	0.275	–	–
32	0.410	–	–
33	0.405	–	–
34	0.315	–	–
35	0.175	–	–

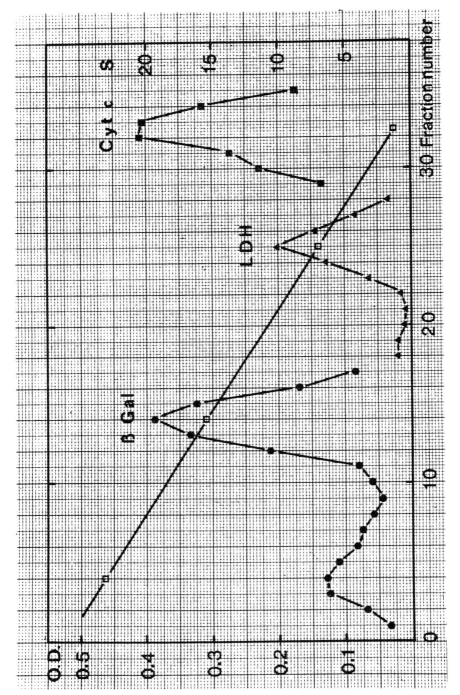

Position of the protein bands in the sucrose gradient and graphic determination of the s-values (□)

Experiment 6 (DNA Isolation)

Total volume of the DNA solution recovered after extraction:
About 9 ml. $O.D._{260}$ of the DNA solution diluted 1:10 = 0.423.

Hence the undiluted DNA solution contains:

$\quad$ 0.423 × 10 = $\quad$ 4.23 O.D. units/1 ml $\qquad$ or

$\quad$ 4.23 $\quad$ × $\quad$ 9 = 34.07 O.D. units total.

Conversion of O.D. units into grams of DNA:
34.07 O.D. units $\triangleq$ 34.07 × 50 µg = 1703.5 µg DNA

Total amount of isolated DNA = 1.7 × 10^{-3} g.

$$\frac{\text{gram isolated DNA}}{\text{gram DNA/phage}} = \frac{1.7 \times 10^{-3}}{2.2 \times 10^{-16}} = 0.77 \times 10^{13} \text{ phages.}$$

The isolated DNA therefore corresponds to the amount of DNA of
0.77×10^{13} phages. As we had 1×10^{13} phages at the beginning,
the yield is approx. 77%.

For DNA free of phenol contamination, the ratio of $O.D._{260}/O.D._{270}$
should be about 1.2. If dialysis is not sufficient, this ratio
will be lower, caused by the heavy UV-absorption of the phenol
at 270 nm.

The DNA could be extracted as the sodium salt, too, if the DNA
solution were adjusted to 0.1 M NaCl instead of 1% (= 0.1 M)
potassium acetate before the alcohol precipitation.

Experiment 7 (DNA Denaturation)

1. T_m for *E. coli* DNA   : 89.5°C

   GC content   : 48.5%

   T_m for salmon sperm DNA: 85.5°C

   GC content   : 38.5%

The *E. coli* DNA was isolated according to HILL, 1968. The salmon sperm DNA was a CALBIOCHEM-product (No. 2620). Both GC contents were found to be about 1.5% lower under our experimental conditions than those quoted in the literature. This could be due to an incompletely developed hyperchromical plateau, or to inaccurate instrument readings.

2. Hyperchromicity

   E. coli DNA: $\dfrac{0.337}{0.250} - 1 = 0.349$

   Salmon sperm DNA: $\dfrac{0.356}{0.245} - 1 = 0.455.$

3. Renaturation (within 3.25 hrs)

   E. coli DNA: $\dfrac{(0.377 - 0.278) \times 100}{0.337 - 0.250} = 68\%$

   Salmon sperm DNA: $\dfrac{(0.356 - 0.312) \times 100}{0.356 - 0.245} = 40\%$

Directions for the Instructor of the Course:

In order to reach temperatures of about 100°C it is preferable to heat the cuvettes by means of an oil bath.
Alternatively, the melting temperature of the DNA can be decreased by lowering the Na^+ concentration of the DNA solution. The GC content of the DNA is then calculated according to

$$GC = \frac{T_m - 81.5 - 16.6 \log \left[Na^+ \right]}{0.41}$$ (HILL, L.R., 1968).

$\left[Na^+ \right]$ means concentration of sodium ions in mole/l.
Needle-shaped temperature feelers with digital recorders (TM15) can be obtained from several suppliers e.g., from METTLER Company, Gießen (FRG).

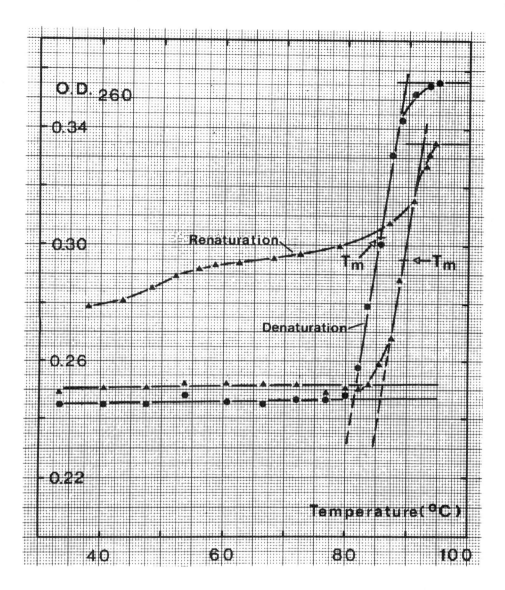

Absorption (O.D.$_{260}$) of DNA solutions in SSC as a function of the temperature, T (°C). (▲) DNA from *E. coli* ; (●) DNA from salmon sperm

Experiment 8 (Isolation of mRNA)

1. From the distribution of the radioactivity and the $O.D._{260}$ readings (see results on pp. 196 and 197), it can be seen that most of the RNA is found in fractions 9-14. The fractions higher than 20 also have high $O.D._{260}$ values. However, the ratio of $O.D._{260}/O.D._{270}$ is <1 and a corresponding maximum of radioactivity is lacking. Therefore these latter fractions only contain phenol. The total amount of RNA in fractions 9-14 is:

$3.130 \times 2.5 \times 10 \times 42 \times 10^{-3} = \underline{3.29 \text{ mg RNA}}.$

2. The specific activity of the RNA is calculated as follows:

Radioactivity of the peak fraction (No. 11), diluted 10^{-1}:

26,585 cpm/20 µl and hence

Radioactivity of the same fraction, undiluted:

$26,585 \times 50 \times 10 = 13.3 \times 10^6 \text{ cpm/ml}$
$= 26.6 \times 10^6 \text{ dpm/ml}$

$$\frac{26.6 \times 10^6}{2.2 \times 10^9} = 12 \times 10^{-3} \text{ mCi/ml}.$$

RNA content of the peak fraction (No. 11), diluted 10^{-1}:

$O.D._{260} = 1.020$ and hence

RNA content of the same fraction, undiluted:

$1.020 \times 10 \times 42 \times 10^{-3} = \underline{0.428 \text{ mg/ml}}.$

The specific activity is:

$12 \times 10^{-3} \text{ mCi}/0.428 \text{ mg} = 2.8 \times 10^{-2} \text{ mCi/mg RNA}$
$2.8 \times 10^{-2} \text{ µCi/µg RNA}$

Calculation of the density of radioactive labeling

The molecular weight of a nucleotide is about 350. The molecular weight of DNA and RNA can be given on the basis of "moles of nucleotides", i.e.,

350 µg DNA or RNA $\triangleq$ 1 µmole.

If the above-calculated specific activity is inserted, the following is obtained:

$350 \times 0.028 \text{ µCi/µmole}$
$9.8 \quad \text{µCi/µmole}.$

According to Avogadro, 6.0×10^{17} molecules correspond to one μmole. Therefore

$$9.8 \text{ μCi correspond to } 6.0 \times 10^{17} \text{ molecules} \quad \text{or}$$
$$(2.2 \times 10^6) \times 9.8 \text{ dpm per} \qquad 6.0 \times 10^{17} \text{ molecules.}$$

Thus, 2.2×10^7 hydrogen atoms will decay per minute among 6×10^{17} nucleotides. Assume that each nucleotide contains only one ^3H atom, then at any given time one ^3H atom would decay per minute out of

$$\frac{6 \times 10^{17}}{2.2 \times 10^7} = 2.7 \times 10^{10} \text{ nucleotides.}$$

The actual density of radioactive labeling is calculated as follows: The half-life of the ^3H is 12.36 years $= 6.3 \times 10^6$ min. Therefore, 1 μmole of this RNA contains $(2.2 \times 10^7) \times (6.3 \times 10^6) \times 2 \approx 2.7 \times 10^{14}$ radioactively marked nucleotides among a total of 6×10^{17} nucleotides. Under the above assumptions, therefore, every 2,200th nucleotide ($= 6 \times 10^{17}/2.7 \times 10^{14}$) contains a ^3H atom. In the case of ^3H the calculation of the density of radioactive labeling is only of theoretical interest, because tritium has a half-life which, in comparison to the duration of of this test, is very long. If, however, we use ^{32}P which has a half life time of only 14.3 days and which upon decay breaks down the DNA or RNA molecules it is important to know the average distribution of radioactivity within a nucleic acid molecule.

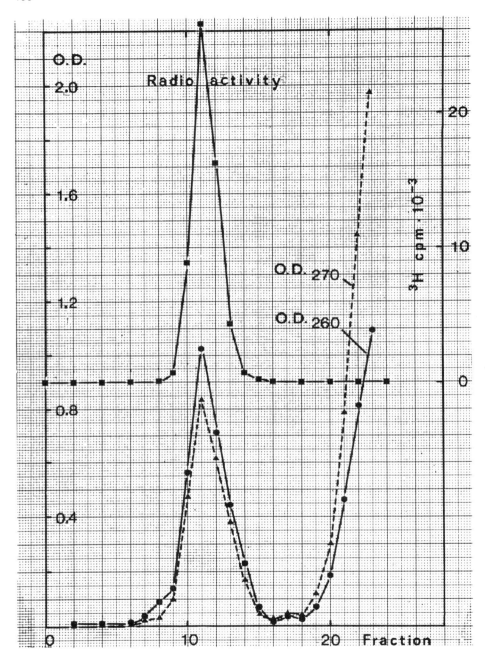

Diagram of elution of the dowex-sephadex-column.
Left ordinate: O.D.$_{260}$ (●) and O.D.$_{270}$(▲). Right Ordinate:
Radioactivity (■)

Experiment 8 (continued)

Fraction No.	O.D.$_{260}$	O.D.$_{270}$	$\dfrac{\text{O.D.}_{260}}{\text{O.D.}_{270}}$	Radioactivity cpm
1	0.002	0.003	∿1.0	46
2	0.002	0.001	∿1.0	29
3	0.003	0.002	∿1.0	38
4	0.002	0.003	∿1.0	31
5	0.003	0.002	∿1.0	32
6	0.003	0.002	∿1.0	40
7	0.005	0.004	∿1.0	42
8	0.091	0.065	1.4	39
9	0.140	0.098	1.4	679
10	0.567	0.480	1.2	8,906
11	1.020	0.840	1.2	26,585
12	0.730	0.618	1.2	16,331
13	0.448	0.384	1.2	4,347
14	0.225	0.170	1.3	769
15	0.075	0.045	1.7	203
16	0.022	0.030	0.7	149
17	0.024	0.045	0.5	162
18	0.023	0.032	0.7	120
19	0.081	0.120	0.7	116
20	0.192	0.306	0.6	118
21	0.467	0.795	0.6	115
22	0.806	1.452	0.6	116
23	1.094	1.960	0.6	135

Incorporation of radioactivity (^{3}H-uridine) into phage mRNA

Balance of ^{3}H uridine incorporation into phage mRNA

	Total volume ml	Sample volume μl	Dil.	cpm/20 μl	cpm/total vol.	Yield %
Total radio-activity 100 μCi	110	20	10^{-2}	378	2.08×10^8	100
Bacterial suspension	3	20	10^{-2}	6,462	9.75×10^7	46.8
Elution of column (fraction 9-14)	15	6×20	10^{-1}	57,617	7.20×10^7	34.6

Experiment 9 (DNA-RNA Hybridization)

When preparing this experiment, T5 mRNA was used for comparison.

Characteristics of the mRNA solutions:

Origin	O.D.$_{260}$	µg RNA/20 µl	Radioactivity cpm/20 µl
T4-infected bacteria	16.3	13.7	108,630
T5-infected bacteria	16.6	13.9	75,150

1. Result of hybridization (corresponds to data sheet II)

Filter	T4 RNA cpm/filter	% hybridiza-tion	T5 RNA cpm/filter	% hybridiza-tion
Control (no DNA)	326 427	0.4	695 659	0.9
T4 DNA	22,832 23,620	21.4	823 775	1.0
T5 DNA	520 423	0.4	15,322 15,118	19.6

Obviously the mRNA molecules of T4 and T5 do not have identical base sequences, because the T4 mRNA radioactivity retained on the T5 DNA filters equals that retained on the control filters. Corresponding results were found when hybridizing T4 DNA and radioactive T5 mRNA.

The radioactive "background" on the filter, i.e., the radio-activity of non-hybridized RNA, can be strongly reduced, if the filters are incubated with 20 µl RNAse (1 mg/ml, DNAse-free) at 37°C for 30 min after hybridization. The RNAse then hydrolyzes all of the single strands, i.e., the RNA molecules which are not completely hybridized.

2. An increase of the amount of RNA would be senseless for two reasons:
a) DNA is transcribed asymmetrically *in vivo* (compare Expt. 8), i.e. at most 50% of the filter-bound DNA is available for hybridization.

b) The mRNA was isolated from the bacteria 5 min after phage infection (compare Expt. 8). At this time, less than half of the phage genome is transcribed (only the "early genes").

If points (a) and (b) are considered together, it must be concluded that at most 25% of the total filter-bound DNA can hybridize with mRNA. Since about 100 μg of DNA were fixed on the filters, no more than 25 μg of RNA can be hybridized. If more RNA were added, the system would be overloaded.

200

Experiment 10 (Phenocopy with Fluorouracil)

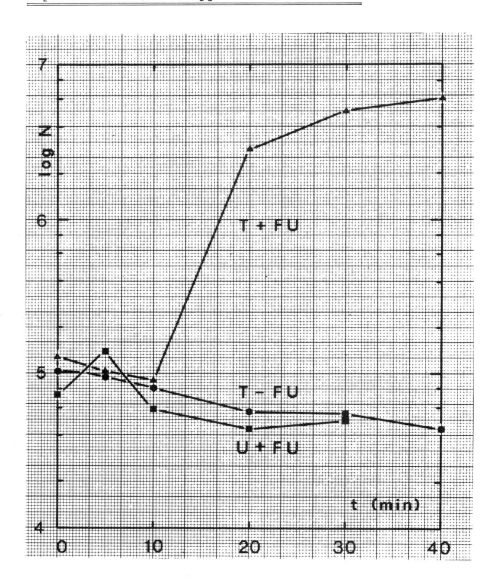

Plaque (N)-formation of T4am N91-infected *E. coli* BA (su) bacteria, which were incubated under different conditions before
plating with indicator CR63 (su$^+$). *t* time of removal of
samples. At 26ºC the transcription of the mutated gene am N91
begins 10 min after infection

Experiment 11 (Genotype and Phenotype)

Data sheet I: S_I = 240 S_{II} = 38
              S_{III} = 255 S_{IV} = 39

Data sheet II: 81.5% prototrophic pig$^+$
               12.9% prototrophic pig
                5.1% auxotrophic pig$^+$
                0.3% auxotrophic pig (contamination?)

16 colonies of the original plates were not transferred, because they were situated on the periphery of the plates.

The colony pattern is first transferred onto M9 agar and then onto NB agar. Since all colonies grow on NB agar these plates serve as a control that all colonies were also transferred onto the preceding M9 plates.

Why are some colonies red at 30°C and colorless at 38°C?

Hypothesis (a) is improbable because it has not been ascertained that certain sequences of single-stranded mRNA are more thermolabile than others.

Hypothesis (b) may be correct. This could be tested by first incubating the bacteria at 30°C (formation of pigment) and afterwards at 38°C. If the pigment were thermolabile, the colonies would become colorless by a shift from 30°C to 38°C. (This is not the case.)

Hypothesis (c) is the most probable one. *In vitro* tests have shown that many enzymes can be converted from a relatively temperature-resistant form to a temperature-labile form by exchange of a single amino acid as a result of mutation (see WILLIAMS et al., 1965).

Directions for the instructor of the course: The original plates were streaked with a mixture of *Serratia marcescens* W225 (wild type), W226 (prodigiosin-negative mutant) and W315 (leucine-requiring mutant). The original plates must be incubated at 37°-38°C for at least 2 days, because otherwise the colonies of W225 and W315 will turn pink to red at room temperature during the course of the experiment.

Experiment 12 (Spontaneous Mutations)

Results of expt.			Expected values (POISSON)		Chi-square		
n	Q_n	$n \times Q_n$	P_n	E_n	Δ_n	Δ_n^2	$\dfrac{\Delta_n^2}{E_n}$
0	135	0	0.412	139	-4	16	0.115
1	123	123	0.369	124	-1	1	0.008
2	65	130	0.162	54.5	+10.5	110.3	2.031
3	11	33	0.048	16.2	-5.2	27.0	1.670
≥4	3	12	0.011	3.7	-0.7	0.5	0.133
	= 337	= 298	= 1.00	= 337.4	= -0.2	Σ = 3.96	

$$\chi_4^2 = 3.96$$

Average frequency of papillae

$$m = \frac{298}{337} = \underline{0.887}$$

Level of significance $\underline{P = 0.40}$

The difference between the test result and the hypothesis of a random distribution of the papillae is therefore not significant.

Spontaneous mutation rate $\alpha = 0.887 \times 0.69/4.2 \times 10^7 =$

$\underline{1.46 \times 10^{-8}}$ per cell per generation.

For the detection of $\underline{his^+}$-cells in papillae-free colonies, several complete colonies are resuspended in 0.3 ml of P-buffer each and then the total of 0.3 ml is plated on M9 agar.

Experiment 13 (UV-Mutagenesis). Chi^2 Analysis

One-Hit Hypothesis

t_i	N_i	M_i	$h_i \pm s_i$ (%)	$N_i t_i$	$p_i = \mu t_i$ (%)	$D_i = h_i - p_i$ (%)	$\dfrac{D_i}{s_i}$	$\dfrac{D_i}{s_i}^2$
1	2,772	3	0.11±0.06	2,772	0.24	-0.13	-2.17	4.71
2	2,036	8	0.39±0.14	4,072	0.48	-0.09	-0.64	0.41
3	2,739	21	0.77±0.17	8,217	0.72	+0.05	0.29	0.08
4	1,519	18	1.19±0.27	6,076	0.96	+0.23	0.85	0.72

$\Sigma = 50$ $\Sigma = 21,137$ $\chi^2_3 = 5.92$

Two-Hit Hypothesis

t_i^2	N_i	M_i	$h_i \pm s_i$ (%)	$N_i t_i^2$	$p_i = \mu^2 t_i^2$ (%)	$D_i = h_i - p_i$ (%)	$\dfrac{D_i}{s_i}$	$\dfrac{D_i}{s_i}^2$
1				2,772	0.08	+0.03	0.50	0.25
4		values see above		8,130	0.34	+0.05	0.36	0.13
9				24,600	0.76	+0.01	0.06	0.004
16				24,300	1.34	-0.15	-0.56	0.31

$\Sigma = 59,802$ $\chi^2_3 = 0.694$

P = 0.11 for one-hit hypothesis;

P = 0.88 for two-hit hypothesis.

To Point (2): Inactivation constant

$$k_{W225} = \frac{2.30(7.81 - 5.52)}{4} = \frac{5.27}{4} = 1.32 \text{ min}^{-1} \text{ (hcr}^+)$$

$$k_{W227} = \frac{2.30(7.83 - 1.40)}{1} = 14.8 \text{ min}^{-1} \text{ (hcr)}.$$

To Point (3): Fraction of lethal damage reactivated by the host

$$\frac{(14.8 - 1.32) \times 100}{14.8} = \frac{1348}{14.8} = 91\%.$$

The "reactivation sector" as concluded from the graph is 0.87.

To Point (4): The absolute radiation sensitivity of kappa and T2

$$t = \frac{2.30(\log 100 - \log 37)}{k} \quad \text{(general formula)}$$

$t_{W225} = \underline{0.75 \text{ min}}$ $\qquad\qquad$ $t_{W227} = \underline{0.067 \text{ min}}$

$D_{W225} = 0.75 \times 11 \times 60 = \underline{\underline{495 \text{ ergs mm}^{-2}}}$

$D_{W227} = 0.067 \times 11 \times 60 = \underline{\underline{44 \text{ ergs mm}^{-2}}}$

$D_{T2} = \underline{\underline{20 \text{ ergs mm}^{-2}}}$

Conclusion: If UV-treated phage kappa is plated under conditions not allowing host-cell-reactivation, it is almost as UV sensitive as the virulent phage T2 which is not host-cell reactivatable. The factor of 2 by which the sensitivity of T2 and kappa differs can result from the fact that T2 contains approx. 2.5 times more DNA than Kappa.

To Point (5): See also the tables for the Chi^2 evaluation (p. 203).

$\mu = 50/21.137 = 0.24\%$ mutants min^{-1}

$\mu^2 = 50/59.802 = 0.084\%$ mutants min^{-2}

The two-hit hypothesis (P = 0.88) is more likely than the one-hit hypothesis (P = 0.11).

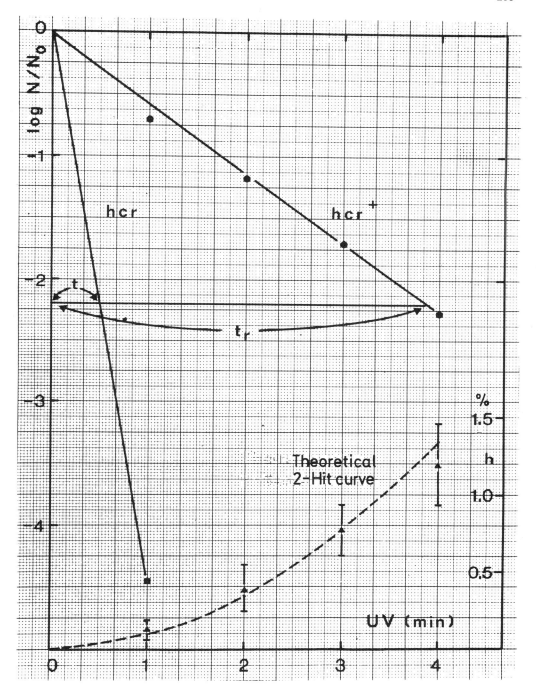

Survival (N/N_o) of phage kappa and the frequency (h) of c
mutants among the survivors as a function of the time of irradiation. Indicators used: <u>hcr</u>[+] and <u>hcr</u>

Experiment 14 (Photochemistry of Cytidylic Acid)

1. Extinction coefficients $[cm^2/mmole]$

The molarity of the Cp-solution was 6.2×10^{-5} M.

Substance	ε_{240}	ε_{254}	ε_{270}
Cp (see p. 208, bottom)	7,600	7,100	9,200 (8,900)
Cp* (see p. 208, bottom)	14,200 (13,000)	6,700	1,130 (1,500)

The numbers in parentheses are taken from K.L. WIRZCHOWSKI and D. SHUGAR, Acta. Biochem. Polonica 8, 219 (1961).

2. Cross section σ for the reaction Cp⟶Cp* (see p. 208, top)

Calculated from the samples irradiated for 0 and 4 min

$$\sigma = \frac{(\log 0.57 - \log 0.26) \times 2.30}{4.0} = \frac{0.314 \times 2.30}{4.0}$$

$$= \underline{0.20} \left[\frac{cm^2}{\mu E}\right].$$

3. Quantum yield Φ

$$\Phi = \frac{\sigma}{2.30 \times 10^{-3} \times \varepsilon_{254}} = \frac{0.20}{2.30 \times 10^{-3} \times 7,100}$$

$$= \underline{0.012} \left[\frac{\mu mole}{\mu E}\right].$$

H. BECKER et al., Photochem. Photobiol. 6, 733 (1967): Irradiating a Cp3'-solution at pH 8.0 with monochromatic UV ($\lambda = 280$ nm), the values found were $\sigma = 0.20$ and $\Phi = 0.022$.

4. After 6 hrs about 90% of the photoproduct was reverted (see p. 208, top):

$$\frac{0.51 \times 100}{0.57} = 90\%.$$

5. Estimation of the photon flux D under our irradiation conditions:

a) At a distance of 10.4 cm (r_1) from the UV-lamp, phage T4 was inactivated to a survival of 0.1% in 5.1 sec. This corresponds to 6.9 lethal hits. According to the literature, 48 ergs mm^{-2} are necessary for 1 lethal hit. This means that at a distance of 10.4 cm the photon flux rate is:

$$F_1 = \frac{6.9 \times 48}{5.1} = \underline{64.9 \text{ ergs } mm^{-2} \text{ sec}^{-1}}.$$

b) To a first approximation the relation between photon flux rate and distance from the source of radiation is given by the equation

$$F_2 : F_1 = r_1^2 : r_2^2 .$$

If the cytidylic acid is irradiated at a distance (r_2) of 3.0 cm from the UV lamp, one obtains

$$F_2 = \frac{r_1^2 \times F_1}{r_2^2} = \frac{10.4^2 \times 64.9}{3^2} = 780 \text{ ergs mm}^{-2} \text{ sec}^{-1}$$

$$= 7.80 \times 10^4 \text{ ergs cm}^{-2}\text{sec}^{-1}$$

$$= \underline{\underline{4.68 \times 10^6 \text{ ergs cm}^{-2}\text{min}^{-1}}}$$

c) 1 micro-Einstein (μE) = 6.02×10^{17} photons

   1 erg $\triangleq 1.28 \times 10^{11}$ photons at a wave length of $\lambda = 254$ nm*.

The photon flux, D, then is

$$D = \frac{(4.68 \times 10^6) \times (1.28 \times 10^{11})}{(6.02 \times 10^{17})}$$

$$\underline{\underline{D = 1.0 \ \mu E/cm^2}} .$$

The value, D, can be determined more accurately, e.g., by actinometry with a solution of malachite-green-leucocyanide. Ultraviolet light (λ = 225 to 290 nm) splits this colorless substance with a quantum yield of Φ = 0.9 (for more details, s. JOHNS, 1969).

---

* This results from the definition of energy
  $E = h \times c/\lambda$ [erg].
  h = Planck's constant 6.62×10^{-27} [erg sec].
  c = Speed of light 3.00×10^{10} [cm sec^{-1}].
  λ = Wave length (cm), e.g., 254 nm = 2.54×10^{-5} cm.
  1 nm = 10^{-7} cm.

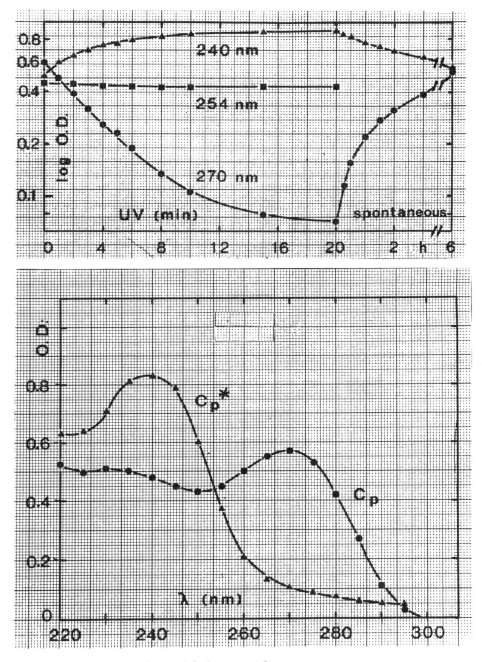

Photochemistry of cytidylic acid

Experiment 15 (Chemical Mutagenesis)

1. Forward mutation

- Survival of the bacteria immediately after 30 min of NG treatment $1.3 \times 10^8 / 2.5 \times 10^8 \approx 50\%$.

- Frequency (%) of mutants

t	Auxotrophs	Mixed clones	Pure clones of pigmentation mutants
0'	0/350	0/350	0/350
30' (without)	0.8[a]	4.9 ± 1.0	0.6[a]
30' (with)	2.3 ± 0.4	0/1,170	4.4 ± 0.6

[a] The number of colonies was to small for the determination of the standard deviation with the nomogram (see page 12).

- Question. Colonies with pigment-sectors were found frequently (4.9%) when the cells were plated immediately after the NG-treatment. Post-incubation of the mutagen-treated cells lowered the fraction of these mixed clones and caused a marked increase of the number of pure mutant clones. This can be interpreted in different ways:

a) The exponentially growing culture contained some pairs of incompletely separated cells. This occurs in many bacterial strains.
b) The cell suspension used for mutagen treatment contained some cell-aggregates, perhaps as a result of the preceding centrifugation.
c) Most of the exponentially growing cells contained 2-4 nucleoides per cell, which is usual.

When exposing a few cells (e.g. a cell clump) to a mutagen, usually only one if any of the cells will mutate. Similarly, only one nucleoide in a multinucleated cell will mutate at the most. When the cells are plated immediately after the action of the mutagen, the result in both cases would be a mixed "clone". A pig mutation might be seen as a colony with colored sectors, but a colony containing auxotrophic cells plus wild type would rarely be detected by the replica technique. On the other hand, division of the mutagen-treated cells during growth in broth would separate mutated and non-mutated cells or nucleoides, respectively (segregation). Thus, the frequency of auxotrophic mutants would seemingly increase during the incubation. This is in fact observed. A method which avoids the difficulty of detecting auxotrophs in mixed colonies has been developed by W. MESSER and W. VIELMETTER, Biochem. Biophys. Res. Comm. 21, 182-186 (1965).

2. Back-mutation

Plate	A	B	C	D
1	500-1,000	–	500-1,000	–
2	approx. 100	approx. 100	approx. 100	–
3	–	./.	./.	–

According to this test, BUdR and NG are mutagenic, but not HA. However, HA is well known as a strong mutagen, specifically attacking cytosine. The discrepancy between our experimental result and the expectation can be explained by the hypothesis that the auxotrophic strain W366 probably does not contain a GC pair at the site of mutation.

BUdR is a pyrimidine which acts as a mutagen only if it is incorporated into replicating DNA. Therefore it is understandable that deoxythymidine competitively interferes with the mutagenic effect of BUdR, but not deoxyadenosine. The mutagenic effect of NG is influenced neither by deoxythymidine nor by deoxyadenosine.

High concentrations of NG and HA inhibit the bacterial "background growth" (M9s instead of M9!) on the agar plates. BUdR does not significantly inhibit bacterial growth. Growth inhibition and mutagenicity of the substances investigated show no correlation.

Experiment 16 (Bacterial Conjugation)

1. 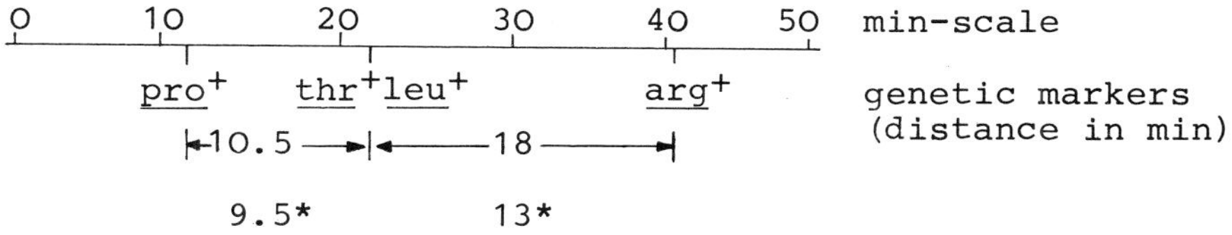
0
10
20
30
40
50
min-scale
pro+
thr+leu+
arg+
genetic markers
(distance in min)
10.5
18
9.5*
13*

* Data from TAYLOR and TROTTER, Bacteriol. Rev. 31, 332-353 (1967).

2. The ratio of Hfr to F^- cells in the conjugation mixture was 1:9, i.e., only 10% of all cells were of the Hfr type. The surplus of F^- cells should make sure, that each Hfr cell will find an F^- conjugation partner.

3. Frequency of recombinants:

$$\frac{340}{(3\times10^{8})\times10^{-1}\times(3\times10^{-1})\times10^{-1}\times10^{-1}\times10^{-1}} = \frac{3.4\times10^{2}}{9\times10^{3}} \approx \underline{\underline{4\times10^{-2}}}.$$

This means that at least 4% of the Hfr cells formed conjugation pairs with the F^- cells, resulting in *pro*$^+$ recombinants.

212

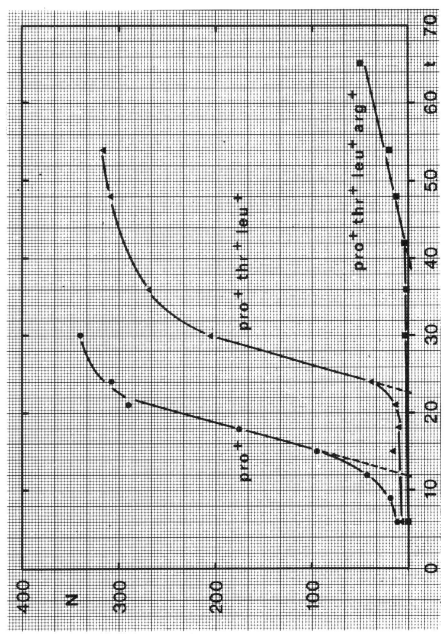

Time-course of the formation of different recombinants by the conjugation Hfr GY767 ×
F⁻AB1157. N = number of colonies per plate. t = time of removal of samples (in min)
after start of conjugation

Experiment 17 (Phage Cross)

Cross	(1)		(2)	(3)	(4)	(5)	
	N_1	N_2				su^+	su
x × y	$7.9×10^7$	$3.3×10^9$	4.79	ca. 12	22	−	−
y × z	$6.4×10^7$	$4.0×10^9$	3.20	ca. 12	27	47%	13%
x × z	$1.7×10^8$	$5.0×10^9$	6.80	ca. 12	33	62%	15%
xz × y	$9.8×10^6$	$2.6×10^9$	0.75	ca. 12	17	−	−

To Question (2): The gene map is

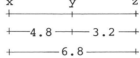

The sum of the single distances (4.8 + 3.2 = 8.0) is somewhat larger than the distance between the outside markers (6.8); however, the result of the three-factor cross xz × y = 0.75% is only consistent with the map shown above, because

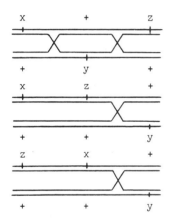

2 crossovers would be necessary; therefore, only few recombinants are to be expected; 0.75% were found.

1 crossover would be sufficient; the frequency of recombinants expected is 3.2% (see cross y × z).

1 crossover would be sufficient; the frequency of recombinants expected is 4.8% (see cross x × y).

To Question (5): As expected, the frequency of the clear plaques on su^+ bacteria (W319) as indicator was close to 50%. The percentage of clear plaques found on su W225 as indicator was only 13 or 15%. There are two possibilities to explain this result:

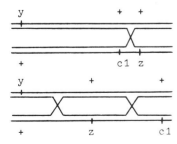

a) The c1 gene is located between sus y and sus z, but very close to the latter.

b) The c1 gene is located on the other side ("to the right") of the sus z gene. A double crossover would be necessary for clear plaque formation on su W225.

Opposing interpretation (a) is the fact that the frequency of clear plaques was found to be very similar in the crosses x × z and y × z. [Further crosses confirmed interpretation (b); U. WINKLER et al., Molec. Gen. Genetics 106, 239-253 (1970)].

To Question (6): The cross xz × y yielded 0.75% recombinants. The expected value for two <u>independent</u> crossovers between x and z is:

$$(4.79\% \times 3.2\%) \triangleq 0.048 \times 0.032 = 0.0015 \triangleq 0.15\%.$$

Then the <u>factor of coincidence</u> would be K = 0.75/0.15 = 5.0. K > 1 indicates "negative interference", i.e., double crossovers occurred more frequently between x and z than was to be expected by random distribution of the crossovers in this region.

To Question (7): $m = m_1 + m_2$ or   12 = 6 + 6

$$P = (1 - e^{-6}) \times (1 - e^{-6}) = (1 - 0.0025)^2 = 0.995$$

i.e., 99.5% of all the bacteria were simultaneously infected by at least one phage of each parental type.

Experiment 18 (Transduction; data sheet I)

Plate No.	Colonies or plaques	Colony or plaque titers	Abbreviations
1 2	169 140	1.55×10^9	Ph
3 4	O O	O	–
5 6	154 140	4.7×10^8	R
7 8	1 1	$1 \quad \times 10^1$	S_L
9 10	O O	O	S_P
11 12 13	131 122 140	1.31×10^3	L
14 15 16	22 19 24	$2.2 \quad \times 10^2$	P
17 18	395 368	3.82×10^8	S
11a 12a 13a 14a 15a 16a	O O O O O O	O O	

1. Multiplicity of Infection

$$m = \frac{0.2 \times (1.55 \times 10^9)}{1.8 \times (4.70 \times 10^8)} = \underline{0.35} \text{ phages/bacterium.}$$

2. Expected frequency of non-infected bacteria:

$$P_0 = e^{-0.35} = \underline{0.71} \ (\hat{=} 71\%).$$

According to this, the titer of non-infected bacteria would be

$$0.71 \times 0.9 \times (4.70 \times 10^8) = \underline{3.00 \times 10^8}/\text{ml.}$$

Thus, the experimentally determined titer of survivors (S = 3.82×10^8/ml) is approx. 27% higher than the theoretically expected titer of non-infected cells. This might be due to the

fact that in the experiment not all of the phages were adsorbed to bacteria.

3. Rate of transduction:

$$\underline{leu}^+ : \frac{(1.31 \times 10^3) - (0.01 \times 10^3)}{0.1 \times (1.55 \times 10^9)} = \underline{\underline{8.4 \times 10^{-6}/phage}}$$

$$\underline{pro}^+ : \frac{2.2 \times 10^2}{0.1 \times (1.55 \times 10^9)} = \underline{\underline{1.4 \times 10^{-6}/phage.}}$$

4. Frequencies of co-transduction (see data sheet II):

	Remarks
$\underline{pro}^+$ with $\underline{leu}^+$: $\frac{0}{381} \leqslant 0.26\%$	no or weak linkage
$\underline{leu}^+$ with $\underline{pro}^+$: $\frac{0}{65} \leqslant 1.5\%$	no or weak linkage
$\underline{ara}^+$ with $\underline{leu}^+$: $\frac{202}{277} = 73\%$	close linkage
$\underline{ara}^+$ with $\underline{pro}^+$: $\frac{0}{65} \leqslant 1.5\%$	no or weak linkage

The calculation of these frequencies is based only on the
colonies which were actually transferred by the replica
technique.

Directions for the instructor of the course: The last host for
replication of phage P1 and therefore the donor was _E. coli_ K12s.

Experiment 19 (Plasmid Transfer)

1. **Elimination of plasmids.** After 12 hrs of incubation, the cell titer in NB without acridine orange was approx. 2×10^8/ml and with acridine orange it was approx. 6×10^7/ml.

Strain	AO	%lac colonies
W1023	–	1/724 = 0.14
W1023	20	162/354 = 45.7
H3000	–	0/420 ≤ 0.2
H3000	20	0/455 ≤ 0.2

The single lac colony, which was found in the random sample of the W1023 culture without acridine orange, may originate from a cell which spontaneously lost the plasmid.

2. Plasmid transfer

Mixture	%lac$^+$ colonies
A	18/326 = 5.52
B	no colonies grown, because of str^s
C	0/582 ≤ 0.2

By incubating the donor and recipient cells together for 15 min, the fraction of lac$^+$ cells among the recipient cells increased by a factor of at least 25.

3. Question. All cells which have become lac$^+$ by plasmid transfer would also be lyso-sensitive for male-specific RNA or DNA phages, which adsorb to the sex pili coded for by the F factor. This lyso-sensitivity could be easily shown in a "spot test" (37°C incubation).

Directions for the instructor of the course: The glassware used for the plasmid transfer experiment must be very clean, since the transfer is sensitive to detergents.

Experiment 20 (Induction of Lysogenic Bacteria)

Bacteria	Ratio	Survival
K12s	$2.5 \times 10^8 / 3.3 \times 10^8$	$= 0.76$
K12(λ)λ^r	$3.8 \times 10^7 / 2.9 \times 10^8$	$= 0.13$
W3110 colD	$9.0 \times 10^6 / 2.0 \times 10^8$	$= 0.045$

Fraction of induced λ lysogenic cells:

$2.4 \times 10^8 / 2.9 \times 10^8 = 0.83$.

Average burst size of the induced λ lysogenic cells:

$(9.3 \times 10^9 - 1.7 \times 10^7)/2.4 \times 10^8 = 39$.

Non-irradiated W3110 colD: 4 AU (corresponds to a 1:   4 dilution).

Irradiated   W3110 colD: 320 AU (corresponds to a 1:320 dilution).

AU = arbitrary units.

By UV induction the extracellular colicin concentration therefore increased by the factor of 80.

Experiment 21 (λ Transfection)

The following λ titers were found in the transfection mixtures:

Tube No.	Titer	ΣPhages/Tube	Efficiency of transfection
1	3.4×10^6	8.8×10^6	2.7×10^{-4}
2a	2.1×10^6	5.4×10^6	1.7×10^{-4}
2b	0	–	–
2c	0	–	–
3	2.5×10^5	6.4×10^5	2.0×10^{-3}
4	4.9×10^4	1.3×10^5	2.9×10^{-3}

1. Proportionality of phage progeny to λ DNA concentration: From the ascending slope of the graph (p. 221, left), it can be seen that a 10-fold increase in the concentration of λ DNA effects a 10-fold higher yield of phages. This direct proportionality (tan α = 1) shows that a single λ DNA molecule is sufficient to transfect a spheroplast. For comparison, drawings have been made on page 221, right, showing the theoretical curves which would be expected if 1 λ DNA molecule (Curve A), 2 molecules (B) or 3 molecules (C) were necessary for the successful transfection of a spheroplast. The tangent of the final linear slope of this curve is 1 (A), 2 (B) or 3 (C), respectively. Curve D would be expected if all the λ DNA molecules were broken into two fragments of lengths a and b and both fragments had to infect a spheroplast simultaneously in order to produce viable phages. The plateau of the curve (p. 221, left) indicates saturation of the "system" with λ DNA, meaning that all of the competent spheroplasts have been transfected so that no further infective centers are formed by the addition of more λ DNA.

2. Efficiency of transfection: See table (above).

3. Frequency of competent spheroplasts:

$$\frac{8.8 \times 10^6}{(4 \times 10^8) \times 250} = \underline{\underline{8.8 \times 10^{-5}}}.$$

The titer of the spheroplasts is 8×10^8/ml; however, only 0.5 ml were added to the transfection mixture.

4a) Questions. Spheroplasts from λ^r cells were used in this experiment in order to prevent the λ phages released from transfected spheroplasts from infecting cells which had not been transformed into spheroplasts. This might increase the phage titer without transfection.

4b) The answer is "no". The λ DNA was so extensively degraded
by the DNAse that there were probably no undamaged molecules
available for the transfection.

Directions for the instructor of the course: The efficiency of
transfection will be further increased if spheroplasts of a
<u>rec</u> B or <u>rec</u> C mutant of *E. coli* are used. For more details and
a method for the production of λ phage DNA, see W. WACKERNAGEL,
Virology <u>48</u>, 94-108 (1972).

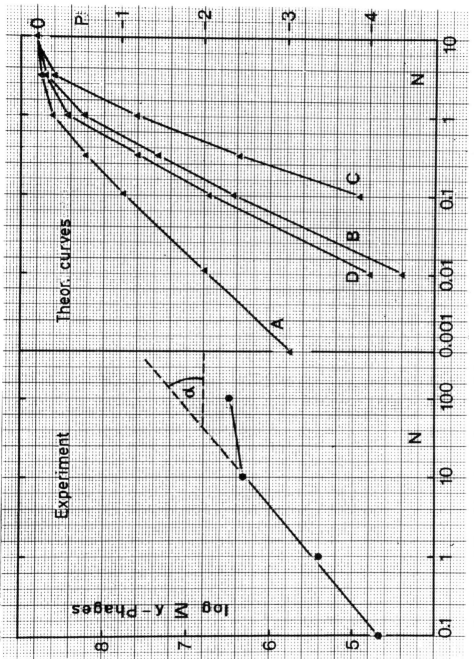

Transfection of *E. coli* spheroplasts with λ DNA. The theoretical curves were calculated according to the POISSON distribution. P = probability for successful transfection. N = number of λ DNA molecules per spheroplast

Experiment 22 (Auxanography and Synthrophy)

Auxanography

Strain	A	B	C	D	Auxotrophic for	Growth halo (diameter)	Background growth
W366	+	-	+	-	Thiamine	3.5 cm, red	++
W626	+	+	-	-	Thymidine	3.5 cm, red	±
W861	+	-	-	+	Histidine	3.5 cm, colorless	±
W982	-	+	-	+	Lysine	3.5 cm, colorless	±

Thiamine is required in a 1,000-fold lower concentration than the amino acids or thymidine because it is a vitamin (vitamin B_1 = aneurine). The heavy background growth of the mutant W366 could have several reasons. Here are a few of them:

- there were traces of thiamine in the agar;

- the W366 culture used for plating contained many prototrophic back-mutants;

- the mutant W366 is "leaky", i.e., the mutation has not entirely abolished the ability to synthesize thiamine; the synthesis then occurs much slower than in the wild type.

Syntrophy

Acceptor	Color of mutant	Inducer 225	592	622	623
W225	red-violet	/	-	-	-
W592	colorless ("white")	±	/	-	-
W622	yellow-orange	-	+	/	+
W623	grey-white	-	++	++	/

The schemes of the biosynthesis of the pigment which could apply are:

The mutants W622 and W623 (in these formal schemes) are inter-
changeable. The correct scheme is the first one:

Precursor B: 2-methyl-3 amylpyrrol
(MAP)

Precursor C: 4-methoxy-2,2'-bipyrrol-
5-carboxaldehyde (MBC)

Pigment: Tripyrrol (see p. 149)

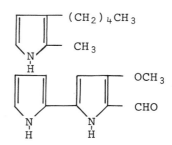

Answers to the questions

- Precursor B (= MAP) is volatile, as W623 is also fed by
  W592 in the double-plate experiment. Precursor C is not
  volatile.

- Wild-type bacteria (W225) do not feed the pig mutants at all,
  or do so only weakly, because the synthesis of prodigiosin
  might be regulated in such a manner that precursors B and C
  (at least) are not accumulated.

- The yellow-orange appearance of the inoculation streak of
  W622 (control plate) can be explained as follows: either a
  precursor of prodigiosin, which is still synthesized by mutant W622, is itself yellow-orange, or if it is colorless,
  it may be converted to a yellow-orange pigment by a shunt
  reaction. The latter is the case. The precursor which accu-
  mulates in strain W622, reacts with precursor B to produce
  norprodigiosin in which the 4-methoxy group of prodigiosin
  is substituted by a hydroxyl group.

Experiment 23 (Complementation)

Complement partner	Complete lysis	Single plaques	Controls
a + b	+	−	a 0 plaques on <u>su</u>
a + c	+	−	b 4 plaques on <u>su</u>
a + d	+	−	c 0 plaques on <u>su</u>
a + e	−	−	d 0 plaques on <u>su</u>
			e 0 Plaques on (<u>su</u>)
b + c	+	−	a lysis on <u>su</u>$^+$
b + d	+	−	b lysis on <u>su</u>$^+$
b + e	+	−	c lysis on <u>su</u>$^+$
c + d	−	31	d lysis on <u>su</u>$^+$
c + e	−	−	e lysis on <u>su</u>$^+$
d + e	−	12	

1. The <u>sus mutants a, b and c</u> are non-allelic.

2. The <u>sus mutants c and d</u> are allelic, but not isogenic, because 31 plaques appeared in the mixture of c and d, although
there were no plaques (= spontaneous back mutants) in the controls of c and d on <u>su</u> W225. The 31 plaques probably represent
wild type recombinants. The same holds true for mutants d and e.

3. The <u>sus mutant e</u> is probably a double mutant, as it comple-
ments neither with a nor with c (e was formed by crossing a × c).

4. Soft agar from a lysed zone (complementation zone), e.g.,
a + b on <u>su</u> bacteria, may be resuspended in a small amount of
sterile buffer and titered after appropriate dilution on <u>su</u>$^+$
and on <u>su</u> bacteria (soft agar method). All phages form plaques
on <u>su</u>$^+$; however, only wild type recombinants and spontaneous
back mutants form plaques on <u>su</u> (by single infection). The
phage titer should normally be much lower on <u>su</u> than on <u>su</u>$^+$.
If this is not the case, wild-type recombinants have been formed
during the complementation.

    a = <u>sus</u> L90   mutated in Gene L
    b = <u>sus</u> N95   mutated in Gene N
    c = <u>sus</u> 086   mutated in Gene O, site No. 86
    d = <u>sus</u> 0191   mutated in Gene O, site No. 191
    e = <u>sus</u> 086 <u>sus</u> L90   mutated in Gene O and L

Experiment 24 (ß-Galactosidase)

Bacterial strain	t min	O.D.$_{580}$	F	ΔO.D.$_{420}$	Δt min	Enzyme Units uncorr.	corr.
Wild type grown with lactose	O	0.21	1.00	O	5	O	O
	10	0.23	0.91	0.03	4	10	9.1
	20	0.26	0.81	0.072	5	19.2	15.5
	30	0.30	0.70	0.096	4	32.0	22.4
R-mutant grown without lactose	5	0.21	1.00	0.33	4	110	110
	35	0.275	0.76	0.23	2	153	116

30 min after induction (addition of substrate) the wild type cells have formed only approx. 20% of the amount of ß-galacto-sidase which was present in the constitutive mutant (R) at an identical cell density

$$\frac{22.4 \times 100}{116} = \underline{\underline{19.3\%}}.$$

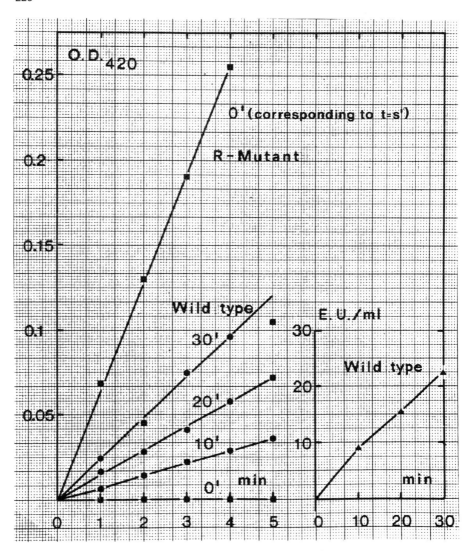

Relative (left) and absolute (right) activities of β-galactosidase in samples taken at different times after the induction of
the bacteria by TMG. *E.U./ml* Enzyme units per ml corrected for
cell densities

Experiment 25   (*in vitro* Morphopoesis)

Data sheet II

	Ratio	Frequency of $\underline{sus}^+$ back mutants at 0 and 200 min
$\underline{am}$ N120	A_0/B_0	$5.6 \times 10^4/1.2 \times 10^8 = 4.7 \times 10^{-4}$
	A_{200}/B_{200}	$5.2 \times 10^4/1.0 \times 10^8 = 5.2 \times 10^{-4}$
$\underline{am}$ B17	C_0/D_0	$9.6 \times 10^4/6.4 \times 10^7 = 1.5 \times 10^{-3}$
	C_{200}/D_{200}	$1.0 \times 10^5/6.2 \times 10^7 = 1.6 \times 10^{-3}$
$\underline{am}$ N120	E_0/F_0	$2.0 \times 10^5/8.0 \times 10^7 = 2.5 \times 10^{-3}$
$+$	E_{200}/F_{200}	$2.2 \times 10^5/3.1 \times 10^9 = 7.0 \times 10^{-5}$
$\underline{am}$ B17		

$\underline{am}$ N120 $+$ $\underline{am}$ B17	Ratio	Frequency of infectious phages formed by *in vitro* morphopoesis
Indicator CR63	$\dfrac{2 \times F_0}{(B_0+D_0)}$	$\dfrac{2 \times 8.0 \times 10^7}{(1.2 \times 10^8 + 6.4 \times 10^7)} = 0.87$
	$\dfrac{2 \times F_{200}}{(B_{200}+D_{200})}$	$\dfrac{2 \times 3.1 \times 10^9}{(1.0 \times 10^8 + 6.2 \times 10^7)} = 38.7$
Indicator BA	$\dfrac{2 \times E_0}{(A_0+C_0)}$	$\dfrac{2 \times 2.0 \times 10^5}{(5.6 \times 10^4 + 9.6 \times 10^4)} = 2.6$
	$\dfrac{2 \times E_{200}}{(A_{200}+C_{200})}$	$\dfrac{2 \times 2.2 \times 10^5}{(5.2 \times 10^4 + 10 \times 10^4)} = 2.9$

Production of the cell extracts for the *in vitro* morphopoesis:
The cultures of phage-infected cells were concentrated by
centrifugation to a cell density of

$$200 \text{ ml} \times (4 \times 10^8) = 8 \times 10^{10}/\text{ml}.$$

Assuming that

- all cells were phage-infected,
- 99% of these cells were disintegrated by freezing and thawing and
- each cell produced approx. 20 phage heads or phage tails alternatively,

each extract should have contained

$$0.99 \times (8 \times 10^{10}) \times 20 \approx 1.6 \times 10^{12}/\text{ml}$$

phage equivalent substructures. In addition to this, however,
the extract also contained infectious phages, as proven by
plating on indicator $\underline{su}^+$CR63. Example:

$$\text{T4 } \underline{am} \text{ N120} = \frac{1.2 \times 10^8 \text{ phages}}{1.6 \times 10^{12} \text{ equivalents}} \approx 10^{-4}.$$

As a correspondingly high phage titer was not found on indicator
su BA, these infectious phages must come from the "extracts"
of the sus mutants. Possible origin:

- Phages which did not adsorb during the infection of the
  original bacterial culture on su BA but were "trapped" in
  the sediment after centrifugation.

- Progeny of such sus phages which adsorbed on su BA and
  multiplied as a result of transmission.

In vitro morphopoesis was evident: When the mixture of am N120
and am B17 extracts was incubated, the phage titer on indicator
su⁺CR63 increased to a value 39 times higher than the control.
As a corresponding increase of the plaque titer on indicator
su BA was not found, the phages formed by extracellular morphopoesis must be sus mutants.

Directions for the instructor of the course: The given incubation time of 200 min can be shortened to 100 min with no significant loss.

B. Answers to the Problems

Concentration of Phages and Ultracentrifugation

1a) The phage forms a repressor. This prevents the transcription of those phage genes which are necessary to produce infectious progeny. Only found with temperent phages.
    Example: λ + $E.\ coli$ K12 ⟶ K12 (λ)

1b) The phage DNA is degraded immediately after infection by a bacterial endonuclease (restriction enzyme). Only possible when phage and host DNA differ in their pattern of methylation ("modification").
    Example: $\lambda \cdot$ K12 m_B + $E.\ coli$ K12m_K ⟶ Restriction

($\lambda \cdot$ K12m_B means that the last host for multiplication was a K12-strain with the modification pattern "B").

1c) A bacterium is immune to lytic phage production when it is lysogenic and the superinfecting phage is homologous to the prophage (compare 1a).
    Example: λ + $E.\ coli$ K12(λ) ⟶ no phage production.

1d) Non-permissive conditions exist when

- the infecting phage is a nonsense mutant and the host has no suitable active suppressor (tRNA).
  Example: T4amber + $E.\ coli$ BA(su)

- the infecting phage is a temperature-sensitive (ts) mutant, and the infected bacteria are incubated at 40° instead of 30°.
  Example: T4 ts.

2) Characteristics of the nucleic acids of different phages

Phage	lin.	circular	single str.	double str.	term. redundancy	cyclic permutation	cohesive ends
T2, T4	+	-	-	+	+	+	-
T3, T7	+	-	-	+	+	-	-
λextracellular	+	-	-	+	-	-	+
λintracellular	-	+	-	+	-	-	-
ΦX174, fd extracellular	-	+	+	-	-	-	-
intracellular	-	+	-	+	-	-	-

3a) Isopycnic (equilibrium) gradient centrifugation to determine
the buoyant density. At t = 0 the molecules to be investigated
and the molecules which form the density gradient (e.g., CsCl),
are evenly distributed. At t = 1 the band has the identical
position as at t = 1 - x.

3b) Zonal-centrifugation to determine the sedimentation constants.
The material to be investigated is layered on a preformed density gradient, e.g. of sucrose, at t = 0. During centrifugation
the band migrates until it reaches the bottom of the centrifuge
tube.

Nucleic Acids and Transcription

1) dAMP 331.2
    dTMP 322.2   Average = 1307.8/4
    dGMP 347.2   = 327 daltons
    dCMP 307.2

2) $130 \times 10^6/6.02 \times 10^{23} = 2.16 \times 10^{-16}$ grams/molecule

3) $1.7 \times 10^6/327 =$ 5,200 nucleotides
    $5,200/3 \times 250 =$ 7 proteins

4) Molecular Weight $= \dfrac{a \times b \times c}{d} = \dfrac{17.2 \times 10^4 \times (2 \times 327)}{3.4}$

$$= 33.1 \times 10^6 \text{ daltons}$$

   a = Length of the DNA molecule (in μm)
   b = Factor for the conversion of μm to Å
   c = molecular weight of a nucleotide pair
   d = "length" of a nucleotide pair (in Å)

5a) A total of 15 ^{15}N instead of ^{14}N

   $15 \times 100/1,307.8 =$ 1.15%

5b) A total of 39 ^{13}C instead of ^{12}C

   $39 \times 100/1,307.8 =$ 2.99%

5c) A total of 57 ^{2}H instead of ^{1}H

   $57 \times 100/1,307.8 =$ 4.36%

5d) MW of -Br = 79.9 $64.9 \times 100/1,307.8 =$ 4.96%
    MW of $-CH_3$ = 15.0

    Difference = 64.9

6) The sequence can be composed of approx. 11 base pairs, because

   $4^x = 3.5 \times 10^6$ or $4 = \sqrt[x]{3.5 \times 10^6}$

   $x =$ 6.54/0.60 = 10.9

7a) 1,400 × 3 = 4,200 base pairs
     164 × 3 =   492 base pairs
      77 × 1 =   77 base pairs

Mutation and Photobiology

1) Base	BUdR	HNO_2	NH_2OH	2AP	UV
A	−	+	−	+	−
T	+	−	−	−	+
G	−	+	−	+	−
C	+	+	+	−	+
	Substitutes	Deamination Cross linking	Ring cleavage	Substitutes	Dimerization Formation of hydrate Deamination Cross linking

2) Mutant	Wild type	Mutant	Type of change[a]	Number of possibilities
1	UUU UUC	UUA UUG	Transversion in 3rd nucleotide	4
	UUU UUC	CUU CUC	Transition in the 1st nucleotide	2
2	UUU UUC	UGU UGC	Transversion in 2nd nucleotide	2
3[b]	UGG	UGA UAG	Transition in 12th or 11th nucleotide, respectively	2
4	UGG	GGG	Transversion in 10th nucleotide	1
		CCU GGX CCX GGX	Deletion of 9th or 10th nucleotide, respectively	2

[a] The sequence of 4 amino acids corresponds to a continuously numbered sequence of 12 nucleotides. A base position which can be filled by any base is symbolized by X.

[b] The polypeptide fragment of mutant 3 can only be synthesized in the absence of an amber or opal suppressor.

(Wild type)   (Mutant 5)

phe ser pro trp ──────→ phe ser ala leu
 | | | | | | | |
UU^U_C UCX CCX UGG ──────→ UU^U_C UCX GCX CUX

or UU^U_C AG^U_C CCX UGG ──────→ UU^U_C AG^U_C GCX UU^A_G

1st Possibility: Insertion of C between 4th and 5th nucleotide

UU^U_C U$\underline{C}$C GCC ^C_UUG

2nd Possibility: Insertion of X between 5th and 6th nucleotide

UU^U_C UC$\underline{X}$ GCC ^C_GUG

3rd Possibility: Insertion of G between 6th and 7th nucleotide

UU^U_C UCX $\underline{G}$CC ^C_GUG

The possibilities 1-3 only exist when ser was coded by UCG and
pro by CCC or CCU.

3) The critical size of the population is reached at
N = 3.5 × 10^5 cells.

Transfer and Recombination of Genetic Material

1) Assuming that the phage genes are not linked linearly, but
in a circular manner, there is no contradiction between the
results of the two and the three factor crosses. Circular linkage maps result when

- the phage DNA has a cyclic structure
  as for example phage ΦX174, or
- when the phages have a linear and terminally
  redundant DNA molecule with a cyclically
  permuted base sequence as, for example,
  T4.

2) Labeling

Parent I : "light", radioactive DNA, e.g. ^{32}PO$_4$$^{---}$
           or ^{3}H thymidine

Parent II: "heavy", non-radioactive DNA, e.g.
           5-bromodeoxyuridine or ^{15}N and D$_2$O.

The DNA must be centrifuged in a CsCl density gradient.
Hybrid DNA is radioactive and medium-heavy. Joint molecules,
in contrast to recombinant molecules, will separate into both
of their parental components by heat denaturation. ("light"
and "heavy" refer to the buoyant density of the DNA.)

Phenotypic Expression

1a) 4^2 = 16 different amino acids (doublet code)

1b) 4^3 = 64 different amino acids (triplet code)

2) Eight different triplets (2^3 = 8) are formed:

Triplet	Chance for formation	Amino acids
AAA	$0.7 \times 0.7 \times 0.7 = 0.343$	lys
AAC	$0.7 \times 0.7 \times 0.3 = 0.147$	asn
ACA	$0.7 \times 0.3 \times 0.7 = 0.147$	thr
CAA	$0.3 \times 0.7 \times 0.7 = 0.147$	gln
ACC	$0.7 \times 0.3 \times 0.3 = 0.063$	thr
CCA	$0.3 \times 0.3 \times 0.7 = 0.063$	pro
CAC	$0.3 \times 0.7 \times 0.3 = 0.063$	his
CCC	$0.3 \times 0.3 \times 0.3 = 0.027$	pro

$$\text{Sum} = 1.000$$

As the genetic code is degenerate, only 6 different amino acids will be coded. The most frequent one would be lysine.

3) The DNA of bacteria rich in alanine and arginine would have to have a higher GC content than the DNA of other bacteria. If this assumption holds true, the bacterial strains under consideration should also contain less isoleucine.

4a) 2,560 nucleotide pairs are incorporated per second, because

$$\frac{3 \times 10^9}{654 \times 30 \text{ min} \times 60 \text{ sec}} = 2.56 \times 10^3.$$

(Average MW of a nucleotide pair = 654).

4b) (1) There are many initiation sites and, (2) there is repeated transcription of the same DNA segments at the same time.

5) Example	Synthesis of protein			
	without inducer		with inducer	
	protein A	protein B	protein A	protein B
1	+	+	+	+
2	-	-	+	+
3	-	-	-	+
4	-	+	+	+

6a) Actinomycin D inhibits transcription by binding to the guanine bases of double-stranded DNA.

6b) Chloramphenicol inhibits protein synthesis by binding to the 50 S subunits of the ribosome.

6c) Mitomycin C inhibits the DNA synthesis. It alkylates mainly guanine bases and produces "cross links" between the two strands of double-stranded DNA. DNA is degraded by an unknown secondary reaction.

6d) Penicillin G inhibits <u>cell wall formation</u> in growing bacteria. Probably a transpeptidase is acylated and thus the formation of the peptido-glycan sacculus is prevented.

6e) Puromycin inhibits <u>translation</u>. It can be incorporated into any growing polypeptide chain instead of amino-acyl-tRNA's. But it is unable to form peptide bonds with the subsequent amino acid. The result is a premature stop in the growth of the poly-peptide chain.

6f) Rifampicin prevents the <u>initiation of transcription</u>. The attachment of rifampicin to the β-subunit of the DNA-dependent RNA polymerase probably renders the initiation complex unstable.

7a) Tryptophane, tyrosine and phenylalanine are aromatic amino acids, which have chorismic acid as a common precursor. If the biosynthesis of chorismic acid is blocked by mutation, the cells are polyauxotrophic for the above amino acids.

7b) Point mutants which have lost the ability to utilize several sugars due to a single mutation frequently lack

- a protein necessary for the transport of the different sugars into the interior of the cell <u>or</u>

- there is no cyclic 3',5'-adenosine monophosphate (cAMP) or "CAP" (cAMP Acceptor Protein) available; both are necessary for the substrate-induction of the different "sugar operons".

The Genetic Code

The 64 possible mRNA base triplets and the amino acids which are coded by them:

1st Base (5' end)	2nd Base U	C	A	G	3rd Base (3' end)
U	PHE	SER	TYR	CYS	U
	PHE	SER	TYR	CYS	C
	LEU	SER	N2,ochre	N3, opal	A
	LEU	SER	N1,amber	TRP	G
C	LEU	PRO	HIS	ARG	U
	LEU	PRO	HIS	ARG	C
	LEU	PRO	GLN	ARG	A
	LEU	PRO	GLN	ARG	G
A	ILE	THR	ASN	SER	U
	ILE	THR	ASN	SER	C
	ILE	THR	LYS	ARG	A
	MET	THR	LYS	ARG	G
G	VAL	ALA	ASP	GLY	U
	VAL	ALA	ASP	GLY	C
	VAL	ALA	GLU	GLY	A
	VAL	ALA	GLU	GLY	G

N1, N2 and N3 are nonsense triplets; they do not code for amino acids, except in the presence of suppressing tRNA. As a result of suppression, for example, the triplet N1 can code for serine, tyrosine or glutamine.

AUG and GUG are also starter triplets for the translation when they are located at the beginning of an mRNA.

Subject Index

R. Rieger
A. Michaelis
M. M. Green

A Glossary of Genetics and Cytogenetics

Classical and Molecular

Third completely revised edition

The aim of the book is to list, explain or define, and collate the special terminology which has grown up over the past 100 years in the fields of genetics, cytology and cytogenetics. Some 2500 terms have been catalogued in alphabetical order, their originator named, their synonymity with related terms given, and their meaning explained in detail. Thus the text is not simply a dictionary of terms but rather a short encyclopedia designed to give the reader a thorough understanding of each term's significance and relevance. This glossary is most useful to both specialists and non-specialists. The terminology ranges from expressions of merely historical interest to the latest developments in molecular genetics and biology.

Springer-Verlag Berlin Heidelberg New York

The Ribonucleic Acids

Edited by
P. R. Stewart and **D. S. Letham**

This textbook sets out to close the gap between the rather general approach of biochemistry texts and the bewildering diversity of research reports on the role of RNA in living cells. It presents modern concepts of cell differentiation from the point of view of the processes going on at molecular level, and the key role of RNA in the transfer and decoding of genetic information.

Established biochemists and microbiologists will find this work a help in updating their knowledge, and it will be of considerable interest to workers in virology, cell biology, and genetics.

Other biologists, organic chemists, and biophysists could use the book as an introduction to a new field.

Contents
D. S. Letham, P. R. Stewart, and G. D. Clark-Walker: RNA in Retrospect – G. M. Polya: Transcription. – H. Naora: Nuclear RNA. – A. J. Howells: Messenger RNA. – D. S. Letham: Transfer RNA and Cytokinins. – L. Dalgarno and J. Shine: Ribosomal RNA. – G. D. Clark-Walker: Translation of Messenger RNA. – P. R. Stewart: Inhibitors of Translation. – P. R. Stewart: Mitochondrial RNA. – P. R. Whitfeld: Chloroplast RNA. – A. J. Gibbs and J. J. Skehel: Viral RNA. – R. Poulson: Isolation, Purification and Fractionation of RNA. – Subject Index.

Springer-Verlag Berlin Heidelberg New York